THE VERMONT
WEATHER BOOK

THE VERMONT WEATHER BOOK

by David M. Ludlum

Vermont Historical Society
Montpelier, Vermont
1985

© 1985, 1996 by the Vermont Historical Society, Inc.

All rights reserved. First edition 1985
Second edition 1996

Printed in the United States of America

99 98 97 96   7 5 4 3

Library of Congress Cataloging-in-Publication Data

Ludlum, David McWilliams, 1910–
 The Vermont weather book.

 Bibliography: p.
 1. Vermont—Climate. I. Title.
QC984.V5L83   1985   551.69743   85-6223
ISBN 0-934720-31-2
ISBN 0-934720-30-4 (pbk.)

ISBN 0-934720-30-4

Glossary from *The American Weather Book* by David Ludlum. © 1982 by David Ludlum.
Reprinted by permission of Houghton Mifflin Company. All rights reserved.

PREFACE
TO THE SECOND EDITION

Lhe world continues to spin on its axis, constantly bringing new seasons of new years. Weather occurs every day, sometimes record-breaking but more often close to the average. For this second edition of *The Vermont Weather Book,* I compiled material covering the notable events from 1986 to 1993. As I worked on the new text, however, one of the outstanding weather months of an outstanding winter unfolded. I therefore included in my new compilation a summary of January 1994, since many years may pass before its equal goes into the record books.

Practically all of the new material was reported by the faithful core of official cooperative observers of the National Weather Service. Their daily readings of thermometers and rain gauges form the basis of the climatological record of Vermont. The supervisor of the observers is Dr. Keith L. Eggleston of the Northeast Regional Climate Center, Cornell University. He serves as the regional climatologist for all of New England and the Mid-Atlantic states. I am indebted to his staff, whose diligence provided me with the raw data for the updates. For bringing tables up-to-date, thanks also go to Dr. William E. Minsinger, president of the Blue Hill Meteorological Observatory in East Milton, Massachusetts.

David M. Ludlum

PREFACE
TO THE FIRST EDITION

The weather, quite as much as geography, has shaped Vermont's history.

From the time a severe thunderstorm forced our state's founders to stay in session at Windsor to complete the writing of our first constitution down to the present time when bountiful snowfalls invigorate the ski economy, weather has affected our history and our way of life. Our comparatively short growing season has determined the kinds of crops we can best raise, and the cold nights and warm days of spring make possible Vermont's renowned maple industry.

Vermont's weather is never passive. It lacks the langorous quality of those parts of the nation where one can hardly tell the difference among the seasons. When it's winter in Vermont, you know it. Deep cold, icy travel, and mountainous snows simply can't be avoided. Likewise when the warm breezes of early summer arrive or the crispness of October's "bright blue weather" urges you to keep moving, you react to weather's influences. In a still predominantly rural state, perhaps Vermonters, like their agricultural ancestors, are more attune to the vagaries of the weather than their big city cousins.

There certainly is no subject more frequently discussed by more people than the weather. Without doubt weather reports in the media are among the most popular offerings. Charles Dudley Warner wrote nearly a century ago, "Everybody talks about the weather, but nobody does anything about it."

Well, not quite, because David M. Ludlum has done something about it. For years he has studied meteorology and written books and articles about it. Many know him as the weather columnist for *The Country Journal*, a task he has had since the magazine's founding. Historians and sociologists know him as the author of the frequently cited *Social Ferment in Vermont 1791-1850*, published by the Columbia University Press in 1939. He combines his interest in Vermont and his immense store of meteorological knowledge in this first full-length book ever written about Vermont weather.

It is to this book that Vermonters and their visitors will turn in years to come to learn about their state's weather past and to settle those endless arguments about the hottest day, the year with the most snow, or the number of times Lake Champlain did not freeze over.

Weston A. Cate, Jr., *Director*
Vermont Historical Society

CONTENTS

PART 2

PART 3
HISTORIC STORMS OF VERMONT

September Snows
October Snows

PART 4
WEATHER WATCHING IN VERMONT

VERMONT
WEATHER
AND CLIMATE

THE VERMONT ATMOSPHERE

The atmosphere overhead is Vermont's most valuable possession. Without its nice balance of elements, life on earth would not be possible. The uppermost level, the stratosphere, serves as a protective shield that filters and absorbs deadly cosmic rays radiating from the sun, and also acts as a buffer zone in which meteoric missiles from outer space are atomized. At middle levels, water vapor forms clouds and initiates the precipitation so necessary for the growth of vegetation on earth and for a multitude of human activities. At ground levels the atmosphere provides the proper mixture of chemical gases that vitalize human life and all living organisms.

The winds of the lower atmosphere are constantly at work. They transport huge amounts of heat from the tropics and moisture from the oceans to make our fields and forests thrive and supply the sustenance for living beings. At the same time, the winds and rains working together are steadily reshaping the basic Green Mountain terrain by erosion. Floods and droughts, cold waves and heat waves, gales and calms—all are contributing to a gradual transformation of the Vermont landscape.

The sun has been the ultimate source of the energy that has fashioned the constituents of our atmosphere in a slow evolution ever since the "Big Bang" supposedly started things going in our solar system. The earth's atmosphere was filled with the effluents of fiery volcanos that showered the landscape with fireballs, clouds of dense smoke, dust particles, and lethal gases. Sometime along the course of geologic history, the chemical composition of the air became sufficiently benign to sustain

3

forms of primitive life, and then temperature and humidity conditions modified sufficiently to permit the existence of advanced animal life.

In the most recent geological age, the Pleistocene, the atmosphere was affected by at least four ice ages with intervening periods of optimum warmth. After the withdrawal of the most recent ice sheet about ten to twelve thousand years ago, conditions in Vermont gradually became suitable for human habitation. Just when the first people arrived we do not know with certainty, but climatic conditions probably became sufficiently tempered by 7000 B.C. to encourage the development of a primitive culture by permanent residents.

The atmosphere is still undergoing change. Alternate warm and cool periods lasting several centuries, as well as long spells of wetness and dryness, have resulted in stress to humans and to their agricultural activities. During the past century, the petrochemical age, we have altered the constituents of the very air we breathe, and only in the past three decades have we come to realize the damage we have done. Now we are trying to restore the atmosphere to its former purity. We should not have left to unregulated industrial freedom the composition of our most valuable possession.

Vermont's location on the globe, about equidistant from the equator and the north pole, gives it a more temperate climate than is enjoyed by regions only a few degrees farther north or south. Its median parallel of 44°N determines that the sun will never stand directly overhead as it does south of the Tropic of Cancer, and it will never disappear completely below the horizon as it does north of the Arctic Circle. Though only 1,500 miles from the Tropic of Cancer and from the Arctic Circle, Vermont has an equitable mixture of climatic conditions.

During the cold season, southerly airstreams from the subtropics are able to penetrate northward and bring brief spells of balmy weather as welcome interludes between extended periods of winter cold. In summer, outbreaks of cool air occasionally descend from the polar regions and provide a pleasant relief from the prevailing heat and humidity. Throughout most of the year the alternate flow of airstreams of different thermal and moisture content provides Vermont with a most stimulating environment for human activity.

At noon on the day of the winter solstice, the sun stands at a low of about 21°30' above the southern horizon, while six months later it blazes down from an elevation of about 68°30'. The changing angle of the sun's elevation has important effects on solar intensity and the length of daylight, and these, in turn, control the weather. The intensity of the solar input increases by a ratio of 1 to 4 from midwinter to high summer, and the length of daylight almost doubles from 8 hours and 40 minutes

4

to 15 hours and 38 minutes, making an addition of about 7 hours of direct sunlight per day and an even greater extension of twilight. Each of these factors makes for constant warming as spring and early summer progress. Mean maximum temperatures come about a month after the summer solstice, near the end of July, when temperatures over Vermont average about 68°, up some 50 degrees from the winter nadir in late January.

The mixtures of weather elements making up our Green Mountain weather are carried by airstreams originating in far distant regions—the barren tundras of central and northern Canada; the damp, cool waters of the North Atlantic Ocean; the warm, tropical surfaces of the Gulf of Mexico, the Caribbean Sea, and the tropical North Atlantic Ocean; the hot, dessicated plains of the interior of the Southwest and northern Mexico. On infrequent occasions, air may move all the way from the broad reaches of the North Pacific Ocean on a transcontinental journey of several days. All these airstreams en route to Vermont are modified by the terrain over which they pass and also by mixing with other airstreams along the way. Upon arrival they are further affected by the varied Vermont topography, made up mainly of an elevated plateau eroded into broken, hilly terrain, but still possessing a spine of high mountain peaks flanked by two long valleys.

Since the prevailing atmospheric currents across North America are predominately from the western quadrant, Vermont shares with the rest of the interior a continental type of climate. This is marked by large temperature ranges between summer and winter and by a tendency toward dryness, especially in wintertime. Yet the Atlantic Ocean, lying only about 80 miles southeast of Vermont's eastern border, often extends its maritime influence over western New England. Since the band of heaviest rain or snow lies from 100 to 150 miles to the northwest of a typical storm track, coastal storms moving along the New England seaboard can extend their canopy of precipitation over part or all of the Green Mountain region.

Another oceanic influence is the storm center that stalls off the New England coast for a number of hours or days, usually in the spring and usually in the Gulf of Maine or off Nova Scotia and Newfoundland. It sets off a northeast flow of cool, damp air over northern New England and the adjacent Province of Quebec, and if the chilly winds last long enough, Vermonters speak of a late spring or backward season. Sometimes the spin of winds around a stalled storm center may cause blustery north winds to blow for several days in a row and introduce cold air masses whose influence may continue for a week or more.

In the summer and autumn, tropical storms and even full hurricanes

5

move along the Atlantic seaboard, or occasionally take a land trajectory over the Middle Atlantic States and New England. Moisture evaporated over distant tropical waters and carried northward by a storm system may fall in copious amounts over the Green Mountains, sometimes with disastrous results as in November 1927 and September 1938. The presence of the Atlantic Ocean at no great distance to the southeast and east plays a significant role in the mixture of weather elements making up the total Vermont climate and must be kept in mind by those bold enough to venture a forecast of Vermont's changeable weather.

Basically, the patterns of Vermont's daily weather change because of a never-ending struggle between the atmospheric forces of the north and those of the south. One may dominate for a period of hours or days, eventually to be routed by the opposing force. The north may send a strong high-pressure area southward whose anticyclonic (clockwise) circulation produces a period of fine, settled weather with temperatures running below the seasonal normal. After several days, the cool air at the surface may be overrun by an airstream from the south moving at high elevations. Cloudiness increases gradually, first at upper and then at middle levels, until the entire sky becomes overcast. Soon the cloud layers thicken and lower. A warming trend becomes noticeable. A low-pressure disturbance with a cyclonic, counterclockwise wind circulation may develop to the west and move along the boundary zone between the cool and warm air masses. Precipitation in the form of showers or steady rain or snow soon spreads across the land. The precipitation usually ends with the arrival of a cold front from the northwest, heralding the approach of another high-pressure area and a repetition of the same weather cycle.

The period of dominance of one weather type or the other may vary considerably from cycle to cycle, running from a few hours to a week or more. But change it eventually will. This is the essence of the New England climate, and one may see as many as 136 varieties of weather within a twenty-four-hour period, as Mark Twain with accustomed hyperbole pointed out over a century ago.

On a topographic map Vermont resembles a prehistoric reptile with a high spine and many elevated ribs, which drop to two long valleys running north and south, abruptly on the west and more gradually on the east. This elevated barrier, extending the length of the state from Massachusetts to Canada, splits the state into distinct climatic zones, which show different seasonal patterns and even significant variations in daily weather.

Since the mountain barrier lies astride and at right angles to the normal west-to-east atmospheric flow, airstreams must rise over heights vary-

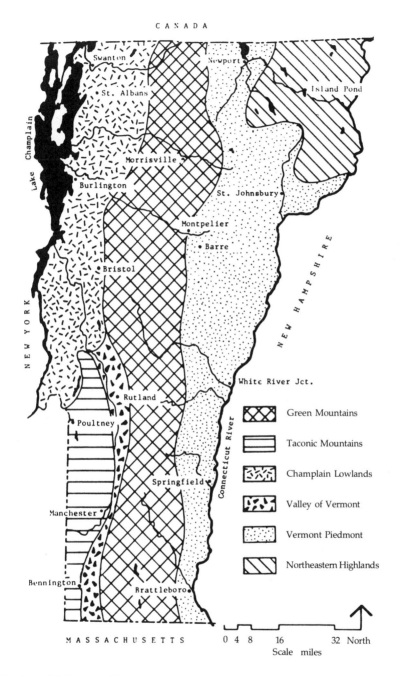

Physiographic Regions of Vermont. From The Nature of Vermont *by Charles Johnson, Used by permission.*

THE VERMONT ATMOSPHERE

ing from 2,000 to 4,000 feet. The ascending air is cooled to its dew point, the upper slopes become cloud-covered, and precipitation often results, with the higher elevations receiving greater amounts than the valleys. This process is quite evident in the precipitation patterns of individual storms. Under a westerly circulation the entire Champlain Valley lies in the rain shadow of the Adirondack Mountains of New York to the west, as does the Connecticut Valley to the east of the Green Mountain massif. When the atmospheric circulation reverses and an easterly flow prevails, the slopes of the mountains to windward experience increased rainfall while the Champlain Valley to the northwest and the Valley of Vermont in the southwest lie in the rain shadow having less precipitation. When a coastal storm moves along the Atlantic seaboard and sets off a northeast flow, the southeast counties on the eastern slope of the Green Mountains often receive much more moisture than do the western or northern counties.

The long finger of Lake Champlain occupies about 100 miles of the western border of Vermont and exerts a tempering influence on the climate of the northwest counties, especially in the autumn and early winter months. Then the warmth of the open waters modifies cross-lake winds carrying an original polar chill. In the Champlain Valley, the arrival of the first frost is thus delayed and the growing season extended, often by as much as thirty days, so crops will mature that cannot withstand the shorter season and earlier freeze of the higher elevations away from the lake.

Lake Champlain freezes over at a relatively late date, usually toward the end of January, and it has been known to remain open all winter in its broadest part. When a cold northwest wind blows over the warmer water in wintertime, it is heated, causing rising convection currents that often generate fog, low clouds, and then precipitation, usually snow squalls over the uplands east of the lake.

Frost hollows are another areal influence on local weather. They occur in Vermont wherever the terrain is uneven. Dense cold air drains by gravity at night from high elevations and becomes trapped in lower, bowl-like areas having no outlets. Frost forms in the bottom of the hollows while the surrounding hillsides may escape the white coating. The surfaces of lakes surrounded by higher ground often fog in during the early morning hours, while the elevations around remain in the clear. Fog is a good protector from frost since it prevents the ground below from cooling to the frost point by further radiation, as any orchard owner along Lake Champlain well knows.

In addition to local valley fog that forms during nighttime radiation, Vermont experiences much fogginess over its higher elevations from the forced ascent of wind flow over a mountain barrier. The upslope motion causes the airstream to cool and condense its invisible water vapor into minute suspended droplets or clouds. This is especially hazardous to aircraft crossing the mountains. Under below-freezing conditions, the forests on the upper slopes become coated with rime ice and highways over the passes may become glazed, making the roads hazardous for automobiles.

The broken terrain of the state and its lofty mountains rising in close proximity give rise to exceptional wind behavior under certain flow conditions. The Valley of Vermont in southwest Bennington County, confined between the Green and Taconic ranges, is subject to destructive local winds, colloquially known as shirkshires. Technically called fall winds or gravity winds, shirkshires flow over the elevated mountain ridge at the head of the valley in Pownal and descend with gathering momentum into the confines of the narrow valley, where a funneling effect increases their speed. They arise when a deep cyclonic storm moving across northern Vermont or southern Quebec induces a strong south or southeast flow. Also southwest winds of great force have been known to sweep the middle elevations of the Green Mountains from Rutland County north and cause immense forest, and even structural, damage. The northeast plateau, too, has been subject to destructive "wind rushes" covering only a local area. These winds appear to originate in mighty downbursts in severe thunderstorms which strike the surface and rush forward in a straight line rather than exhibiting the circular motion of tornadic winds.

Summertime thunderstorms are caused by local conditions when daytime heating of the surface results in rising vertical currents of warmed air. If supplied with sufficient upward impetus by continued heating, the rising columns cool until reaching their dew-point temperature; then condensation takes place and cumulus clouds form. If the updraft from below is of adequate strength and duration and the air contains sufficient moisture, the clouds feed on the released heat of condensation and continue their vertical development to produce swelling clouds or even thunderheads. These drift across the countryside spreading showers and perhaps thunder and lightning along with hail. The mountainous country of Vermont stimulates the creation of vertical currents in the atmosphere and the production of thunderstorms much more readily than does flat terrain.

9

Weather Extremes in Vermont, New England, the United States, North America, and the World

TEMPERATURE (in degrees Fahrenheit)

Maximum

105	Vernon, Windham Co., July 4, 1911
107	Chester and New Bedford, Mass., Aug. 2, 1975
134	Death Valley, Calif., July 10, 1913
136	El Azizia, Libya, Sept. 13, 1922

Minimum

− 50.0	Bloomfield, Essex Co., Dec. 30, 1933
− 69.7	Rogers Pass, Mont., Jan. 20, 1954
− 79.8	Prospect Creek, Alaska, Jan. 23, 1971
− 81.4	Snag, Yukon Terr., Canada, Feb. 3, 1947
− 128.6	Vostok, Antarctica, July 21, 1983

PRECIPITATION (in inches)

Greatest in twenty-four hours

8.77	Somerset, Windham Co., Nov. 3-4, 1927
18.15	Westfield, Mass., Aug. 18-19, 1955
43.00	Alvin, Tex., July 25-26, 1979
73.62	Cilaos, La Réunion, Indian Ocean, Mar. 15-16, 1952

Greatest in one month

16.99	Mays Mills, Windham Co., Oct. 1955
27.70	Torrington, Conn., Aug. 1955
71.54	Helen Mine, Calif., Jan. 1909
88.01	Swanson Bay, B.C., Canada, Nov. 1917
107.00	Puu Kukui, Hawaii, Mar. 1942
366.14	Cherrapunji, Assam, India, July 1861

Greatest in one year

73.61	Mt. Mansfield, Lamoille Co., 1969
130.14	Mt. Washington, N.H., 1969
184.56	Wynoochee Oxbow, Wash., 1931
332.29	MacLeod Harbor, Alaska, 1976
578.00	Puu Kukui, Hawaii, 1950
905.12	Cherrapunji, Assam, India, 1861 (calendar year)
1,041.78	Cherrapunji, Assam, India (season Aug. 1860-July 1861)

Least in one year

22.98 Burlington, Chittenden Co., 1941
21.76 Chatham Light Station, Mass., 1965
  .00 Death Valley, Calif., 1929
  .00 Bagdad, San Bernardino Co., Calif., Aug. 18, 1909 - May 6, 1912 (total of 993 days)
  .00 Arica, Chile, Oct. 1903 - Dec. 1917 (total of 171 months)

SNOWFALL (in inches)

Greatest in twenty-four hours

37.0 Peru, Bennington Co., Mar. 14, 1984
56.0 Randolph, N.H., Nov. 22-23, 1943
75.8 Silver Lake, Boulder Co., Colo., Apr. 14-15, 1921

Single storm

50.0 Readsboro, Bennington Co., Mar. 2-6, 1947
77.0 Pinkham Notch, N.H., Feb. 24-28, 1969
189.0 Mt. Shasta Ski Bowl, Calif., Feb. 13-19, 1959

Calendar month

75.0 Waitsfield, Washington Co., Dec. 1969
78.5 Mt. Mansfield, Lamoille Co., Dec. 1970
130.0 Pinkham Notch, N.H., Feb. 1969
172.8 Mt. Washington, N.H., Feb. 1969
390.0 Tamarack, Calif., Jan. 1911

Season

198.5 Waitsfield, Washington Co., 1970-71
318.6 Mt. Mansfield, Lamoille Co., 1970-71
323.0 Pinkham Notch, N.H., 1969-70
407.6 Mt. Washington, N.H., 1969-70
1,122.0 Paradise Ranger Station, Mt. Rainier, Wash., 1971-72

ATMOSPHERIC PRESSURE (in inches of mercury)

Maximum

31.14 Northfield, Washington Co., Jan. 31, 1920
31.42 Miles City, Mont., Dec. 24, 1983
31.43 Barrow, Alaska, Jan. 3, 1970
31.53 Mayo, Yukon Terr., Canada, Jan. 1, 1974
32.005 Agata, Evenki N.O., Siberia, USSR, Dec. 31, 1968

Minimum

28.28 Burlington, Chittenden Co., Jan. 3, 1913
28.17 Caribou, Maine, Dec. 2, 1942
26.35 Long Key, Fla., Sept. 2, 1935
25.87 Typhoon June, Pacific Ocean, Nov. 19, 1975
       (264 miles west of Guam)
25.69 Typhoon Tip, Pacific Ocean, Oct. 12, 1979
       (520 miles northwest of Guam)

ICE AGE

Vermont is not always green, as year-round residents well know. Mountains and valleys become white each winter, but each spring the snow and ice disappear and fields and forests become green once again. In geologic history not so long ago, the white covering did not go away, but remained throughout the summer. For almost a hundred millennia, the Green Mountain region endured an ice sheet whose remains melted away only some twelve thousand years ago. Known as the Wisconsin Stage, the last of the four great ice advances of the Quaternary Period began about eighty thousand years before the present and reached its maximum development about eighteen thousand years ago. At that time in central Vermont, temperatures on the surface of the ice sheet probably ranged down to − 50°F on winter nights and rose only to the 40°-to-50° range on a midsummer afternoon, as temperatures now do on the ice islands of the Arctic Ocean and along the periphery of the Antarctic ice sheet.

At the height of glaciation the landscape was composed of only one element—ice. A vast sheet of compacted snow overrode all land features as far as the eye could see. There were slight depressions over the present Champlain Valley and the Connecticut Valley, and ridge lines rose over the submerged peaks of the Green Mountains and the White Mountains to the east. All was buried under an almost silent ice sheet, devoid of plant and animal life, whose thickness probably measured between 5,000 and 10,000 feet. The uppermost slopes of both Mt. Mansfield and Mt. Washington show evidence of the ice sheet's presence.

No consensus prevails among paleoclimatologists as to why the climate began to grow colder some 100,000 years ago after a warm interlude (interglacial period). Many hypotheses have been advanced. Suffice it to say here that the surface of the earth received less radiational heat than it does now. As the cold centuries passed, the percentage of precipitation in the form of snow increased and the below-freezing season extended. Eventually, the snow cover remained through the summer, and each winter increased its depth. The snowfields gradually compacted into firn, or granular, snow, and the firn, under the constant pressure from the mass of snow above it, consolidated into ice. Glaciers were born here and there on the Laurentian highlands of interior Quebec and Labrador and grew thicker and denser with passing centuries. The ice mass eventually began to spread out in all directions from the highland centers. Under the colder conditions prevailing, glaciers also formed on the slopes of mountain peaks to the south, and the pressure of the accumulations of ice spread lobes down the slopes into valleys where local ice sheets formed and combined with others.

In a few millennia the unglaciated land in all directions was overridden by the main Laurentian ice sheet in its inexorable southward push. Rock striations and glacial deposits indicate that the ice moved in a southeast direction over the southern sections of Vermont. At its greatest advance the ice sheet reached as far south as Martha's Vineyard, Block Island, and Long Island.

The atmospheric circulation over the vast ice sheet was controlled by a large area of high barometric pressure known as a polar anticyclone, somewhat similar to those existing over Greenland and the Antarctic continent today. With the highest pressure concentrated near the center of the ice sheet and low pressure prevailing around the periphery, a circumpolar whirl of easterly winds prevailed. A distinct continental type of climate existed marked by extreme cold and extreme dryness. Though precipitation was very light, sufficient fell each year to continue the snow-firn-ice process and increase the thickness of the ice sheet. Clear skies and an unpolluted atmosphere, permitting intense nocturnal radiation into outer space, lowered temperatures far below zero at night. Winds were generally light except along the interface of land and ice. There cyclonic storms developed when the cold air from the ice sheet came into contact with warm airstreams from the tropical land and seas. Severe blizzards, like those along the periphery of Antarctica, were generated near and along the edge of the ice sheet.

At some undetermined time and for reasons not yet understood, the atmosphere above the ice sheet began to grow warmer from an input of solar heat, raising temperatures gradually by 5 to 10 degrees F. The

14

arctic refrigerator, which had functioned remarkably well for many thousands of years, changed its thermostatic setting and, like a faulty ice-making machine, failed to cool sufficiently to continue the ice-making process year round. More ice melted in summertime than was added by compacted snow in winter. The circumpolar vortex of winds became less vigorous and gradually contracted into a tighter northerly spiral. This permitted oceanic airstreams to penetrate inland and supplant continental influences along the southern border of the ice sheet. The hard ice began to soften and rot—to ablate, as the geologists call the process. Bare ground began to appear in midsummer, probably first on mountain peaks, while individual glaciers remained in cirques (elevated, bowl-like ravines formed by glaciers) on the upper slopes and in beds of ice in the valleys. In wintertime everything became covered with snow again, but the snow did not compact into ice as before.

The uncovering of Vermont was a very gradual process. Since no terminal moraine marking the leading edge of a glacier has been found within the state, the ice sheet must have melted in place. A glaciologist has calculated that 4,100 years elapsed between the disappearance of the ice in the vicinity of Hartford, Connecticut, and its disappearance at St. Johnsbury in northern Vermont, only 185 miles away. It was not a continuous melting, for temporary climatic swings to colder conditions probably halted the process for a few years or decades. Many isolated bodies of ice remained after the continuous sheet dissolved into fragments.

The receding glaciers often left behind a rocky trail. Frequently the soil was scraped away leaving exposed rock formations. In other cases the land was strewn with rocks which later settlers formed into stone walls. This stone outcropping is in East Wallingford, Vermont.

No one knows exactly when the Green Mountains emerged from the ice sheet and remained bare most of the summer; perhaps it was about fifteen thousand years ago, and perhaps it took another five thousand years for the last vestige of ice to disappear from Vermont. The year 3000 B.C. has been suggested as the date of the final disappearance of the ice remains on the highlands of eastern Canada where it all started.

When the ice finally ablated away, a Vermont terrain much the same as today came into view, but it was quite different from what it had looked like before the ice sheet formed a hundred thousand years before. Soil and forest had been scraped from the tops of the high mountains, leaving bare rock outcrops indented with many striations; some tall hills were completely leveled; great gouges formed depressions in the surface here and there; and many low places were filled with lovely lakes. Each flank of the Green Mountains looked down upon long bodies of water where only narrow rivers had run before.

A closer inspection of the surface revealed many oblong hills (drumlins) and winding mounds of glacial debris (eskers), fashioned by the interaction of ice, meltwater, and ground matter. Here and there large boulders (glacial erratics) were strewn about, which future farmers would encounter with great dismay. Both the Champlain and Connecticut valleys had many elevated terraces (kames), which marked former water levels and would serve in the future as graded paths for highways and railroads.

Lake Vermont

During the immediate postglacial centuries two extensive bodies of water, aligned north and south, occupied sites on either side of the Green Mountains, as do their diminished descendants, Lake Champlain and the Connecticut River. The histories of Lake Vermont and Lake Hitchcock have been adequately described by geologists, but their roles in postglacial Vermont weather require consideration here.

Lake Vermont occupied the preglacial river trough between the Adirondack Mountains of New York and the Green Mountains of Vermont. When the climatic conditions ameliorated during the postglacial era, a large lobe of ice remained in the trough extending between the St. Lawrence Valley and the Hudson Valley. The melt of this ice supplied the waters impounded in Lake Vermont. Originally, the lake drained south into the Hudson Valley where it joined the postglacial waters of Lake Albany. As the land rose, this outlet became blocked and Lake Vermont sought an outlet through the St. Lawrence waterway, which at times was an estuary of the sea. At its widest point Lake Vermont extended from the vicinity of Enosburg Falls on the present Missisquoi River to Harkness, New York, on the Little Ausable River west of Keeseville, a distance of 27 miles, or more than double the present width of Lake Champlain at that latitude. The lake during its early stage stood about 640 feet above sea level, compared to the present height of about 96 feet at the Richelieu River outlet. This relatively large body of water exerted important influences on the climate of Vermont that do not operate today. Since it was several times the mass of the present lake, its response to heating and cooling conditions was more conservative than that of the shallower Lake Champlain of today. The surface froze over later in the season in most years, thus allowing the lake to warm the lands near the shores in late autumn and early winter. In springtime, since the deep body of the water warmed less rapidly than does the present lake, cooling winds from off the lake prevented premature blossoming and damage from late-season frosts. The frost-free season had a much longer span of days near the lake than at a distance back on the surrounding highlands. Further, the prevailing winds from the westerly quadrant were cooled somewhat in their passage over the lake waters in summertime, making for a more temperate season, like Buffalo, New York, now enjoys.

Lake Vermont also served as a local source of moisture under certain meteorological conditions. An outstanding feature of the Great Lakes region climate today is the occurrence of lee-shore showers when the lake water is warmer than the air moving off the surrounding land. Extensive studies of this phenomenon have been carried on by meteorologists who apply the term *lake-effect mechanism* to describe the process that produces the showers. Airstreams from a cool source region, usually to the north or west of a body of water, are heated in the lower layers when passing over the warmer surface waters, creating rising, convective currents and air turbulence in the lower layers of the atmosphere. The ascending air currents carry moisture upward from the lake surface, which cools as it rises and condenses into clouds at the altitude where its dewpoint temperature is reached. Stratocumulus clouds develop at low levels and continue to form as long as the cool airstreams blow across the warm water surface.

Upon reaching the shorelines or the low hills, a short distance inland, added lift is given to the turbulent cloud by the rising terrain, its moisture condenses into water droplets, and a shower results. In late autumn or early winter, when the cool air and warm water have the greatest contrast, the precipitation often falls as snow showers. If the flow of below-freezing air continues, the snowfall may become continuous and accumulate to considerable depths. The flakes have a high ratio of air to

17

moisture and make a light fluffy covering. As much as four feet of snow have fallen in a twenty-four-hour period on the lee shores of Lake Ontario when the lake-effect mechanism was operating in its most proficient manner.

The width of the snowfall belt produced by a lake-effect mechanism may vary from a mile or two to a front of 25 miles, and the depth of its penetration inland depends on the driving energy developed by the winds at snow-cloud level. The direction of the wind trajectory across the lake is important. A northwest wind would have a longer fetch across the lake than a west wind, and a northerly wind might carry the entire north-south length of the lake. The existing weather map situation also is an influential factor. Converging wind flow around a cyclonic storm center is more conducive to the formation of lake-effect showers than diverging airflow around an anticyclonic center.

Even with the present reduced extent and volume of Lake Champlain, the weather station at Enosburg Falls, at an elevation of 442 feet facing the widest part of the lake, has experienced a seasonal snowfall of almost 200 inches, a major portion resulting from snow showers created by the lake-effect mechanism which is in almost daily operation. In the immediate postglacial years, the greater snowfall and its earlier arrival on the western slopes of the Green Mountains would have brought smiles to the owners of ski lodges.

Lake Hitchcock

The Connecticut River Valley trough also experienced an interesting postglacial history. The ablating ice sheet began to uncover the lower valley near Hartford, Connecticut, about 18,000 years ago, and New Hampshire and Vermont ground started to appear perhaps two thousand years later. As the ice lobe melted, its edge retreated northward up the valley, a body of water of considerable size collected in the trough, extending 157 miles from its southern end at Middletown, Connecticut, to its northern end in the Lyme-Thetford area just above Hanover. A natural dam at the southern end prevented the waters from draining into Long Island Sound. The feature was named Lake Hitchcock by later geologists in honor of Edward Hitchcock, a Dartmouth professor and the first state geologist of Vermont. The width of the lake between Vermont and New Hampshire amounted to only two or three miles except where a tributary valley entered and provided an estuary of wider extent.

The imprisoned waters eventually either wore through the blocking dam near Middletown by gradual erosion, or perhaps the sudden shock of an earthquake assisted in unplugging the gorge. When the waters of Lake Hitchcock were released into Long Island Sound and the sea beyond, the level of the lake was supposed to have dropped 90 feet in a few hours.

Lake Hitchcock and its successors probably caused only minor effects on the climate and local weather of the Connecticut Valley when compared to those of Lake Vermont and the Champlain Sea on their surroundings. The lateral width of the bodies of water was much too narrow and restricted and the waters too cold to initiate the lake-effect showers. Also the volume of the water was not great enough to affect local temperatures. No doubt, the presence of cold water along the meadows and intervales induced valley fog to form and create a thermal blanket that served to reduce the danger of radiation frost on late spring and early autumn days. Probably the greatest effect of Lake Hitchcock was hydrologic in augmenting the effect of spring freshets that swept the valley almost annually when the combination of snow melt and heavy rains from a passing cyclonic storm caused mighty floods far exceeding anything witnessed in modern times.

Postglacial Climates

From a variety of direct sources, European climatologists have reconstructed a postglacial chronology of climatic trends for Europe. Contemporary conditions in North America are thought to have paralleled those in northwestern Europe, although the evidence must be supplied by indirect methods.

Following the disappearance of the major portion of the Scandinavian ice sheet, a cool and wet period prevailed for about four millennia from 8000 to 4000 B.C. This was succeeded for the next two thousand years by a "climatic optimum" when temperatures were four to five degrees Fahrenheit warmer than at present. Conditions were drier and less stormy since the main storm belt migrated northward. A deterioration took place from 2000 B.C. to A.D. 400 with a return to a cooler and wetter climate. Then a second warmer and drier period followed for about 800 years, carrying well into the high Middle Ages. The ameliorated conditions permitted the Vikings to rove the North Atlantic in largely ice-free waters as far as the shores of Greenland and Labrador.

A second deterioration covered the two hundred years from A.D. 1200 to 1400, when the permanent arctic ice pack advanced southward to impede sea traffic and cut communications between Europe and Greenland.

A temporary improvement between 1400 and 1550 permitted grain growing to resume in Iceland and northern Scotland, but soon another downward swing of the climatic pendulum brought a return to cooler and wetter conditions during the late sixteenth and seventeenth centuries, the years of the early exploration and first settlement of North America by Europeans. The apt title of Little Ice Age has been employed by paleoclimatologists to describe the period from 1650 to 1890. The most obvious effects of the cooling climate centered in the marked advance of glaciers in Europe and Alaska and changes in the boundaries between boreal and temperate forests.

The years of early settlement in Vermont during the last four decades of the eighteenth century fall within the period of the Little Ice Age, which continued past the middle of the nineteenth century. Temperatures during these years averaged from 1.5 to 2 degrees Fahrenheit cooler than those experienced during the first half of the twentieth century. Frosts came later in the spring and earlier in the autumn than at present, making for a shorter growing season and less abundant crop yields. Pioneer Vermonters had a harder time with the vicissitudes of the daily weather than we do.

An idea of the general climate trend in terms of temperature over the late eighteenth and early nineteenth centuries may be gained from a composite of available temperature records for the eastern United States, which have been reduced to the readings observed at Philadelphia. The record began in 1738 and has been compiled down to the present. Since Philadelphia lies from 3.5 to 5 degrees of latitude south of Vermont, the difference in annual mean temperatures between the two locations amounts to about 8 to 10 degrees Fahrenheit. Yet the two areas lie within the same zone of atmospheric circulation and often share the same type of air mass, witness the passage of the same fronts, and experience similar changes in the daily weather. Thus, the Philadelphia composite record—keeping the temperature differential in mind—may be employed to gain some insight into the long-range climatic trends in the Green Mountain region. The same cannot be applied to the precipitation record, however, since the character, duration, and amount are often affected by local conditions.

The long-term temperature curve for eastern North America, as indicated by the Philadelphia (composite) records, shows two divisions: (1) a mainly cool period encompassing the last part of the Little Ice Age when only brief warm intervals occurred, and (2) a recent period of almost steadily increasing warmth through the 1950s during which only temporary returns to cooler conditions have taken place.

The divide between the two periods came in the middle of the 1880s when the five- and ten-year means of annual temperature began an abrupt rise. The years 1889, 1890, and 1891 were very mild on an annual basis, as were their winters. From 1810 to 1890, all decades, with one exception, averaged below 54.1°F; whereas all subsequent decades from 1890 to the present averaged 54°F or above.

The earliest meteorological record to shed light on actual weather events in Vermont began in 1811 at Williamstown, Massachusetts, where Professor Charles Dewey and his colleagues maintained temperature and precipitation records from 1811 to 1838. Williamstown adjoins Pownal in the extreme southwest corner of Vermont and shares in the same meteorological experiences as that quarter of the state.

During the first cold period of seven years, or eighty-four months, extending from January 1812 to December 1818, only twenty months (24 percent) averaged above the contemporary normal. During this period, only seven months exceeded a + 2 degrees departure from normal, while twenty-eight months had a − 2 degrees or more departure.

In 1812, all months except December were below normal, as was every month in 1817 from January through October. In 1818, January through May ran below, as did August and September. The much publicized "Year Without a Summer" in 1816 showed the greatest departure for the summer season, but was preceded by a mild winter and followed by a pleasant autumn. Cold winters occurred in 1811-12, 1814-15, and 1816-17.

A second cold period extending from 1835 to 1838 appeared prominent in the Philadelphia (composite) records and is locally documented for Vermont by the Williamstown, Massachusetts, records and by a new series started at Hanover, New Hampshire, in 1834. In the latter the three years from 1835 to 1837 numbered among the five coldest in the entire record that extends down to the present, a period of almost 150 years.

The Williamstown record well demonstrates the preponderance of cold months:

In 1835, only October above normal.
In 1836, only January and September above normal.
In 1837, all months below normal.

Climatologists have called attention to the presence of a volcanic dustveil in the upper atmosphere during 1815-16 and again in 1835-36. The dustveil reflects some incoming solar radiation outward, while it allows outgoing terrestrial radiation to pass through unhindered. Each process tends to reduce the amount of heat in the lower atmosphere.

A continuous Burlington record began in 1838 and supplies representative figures for the Champlain Valley down to the present, replacing

21

our reliance on the Williamstown record, which experienced frequent gaps after 1838. Both Burlington and Hanover reached a peak of high temperatures in the graph of running means in 1846 and a trough of low temperatures in 1857.

The second half of the nineteenth century produced alternate periods of above and below normal annual mean temperatures, as had the first half of the century. The cold periods extended from 1855 through 1859, 1866 through 1869, and 1871 through 1876. Similar trends appeared in the Williamstown, Hanover, and Burlington records. Another cold period featured the middle 1880s, with the lowest annual averages coming in 1885 and 1886. The year of Vermont's twin blizzards in January and March of 1888 marked the last year of this cold series.

There is some doubt among climatologists as to whether the Little Ice Age has definitely ended. If it has, the year may be placed at the end of the 1880s, for the decade of the 1890s ran much warmer than any previous one during the century. The winters of 1889, 1890, and 1891 were outstanding for warmth, as were the annual figures. At Hanover, all years from 1894 to 1903 averaged above normal.

Mean annual temperatures again declined during the first twenty-five years of the twentieth century. Burlington reached low points in the running annual means in 1908-09, 1916, and 1924. The outstanding feature of the first half of the century was the rapid warming after 1926, with high points in 1930 and 1951. A minor cold period occurred at the outset of the 1940s, but toward the end of the decade the rise in temperature became abrupt. Both the Burlington and Hanover records show five-year highs centered in 1951 as the warmest period of the entire century. During this period the winters from 1949 to 1953 were noteworthy for mildness.

During the late 1950s the Burlington temperature record began a steady decline, and this continued through the decade of the 1960s. The year 1968 was the coldest calendar year since 1917. A low point in the running means was reached in 1970, but a recovery to higher annual means marked the first half of the 1970s. After 1975, however, a marked decline set in. Each winter from 1976 to 1982 had a spectacular wintry month.

VERMONT WEATHER
MONTH BY MONTH

January

"As the days lengthen, the cold strengthens," goes the familiar post-solstice proverb. With rare exceptions, January is the coldest month of the year. The jet stream, whose meanderings are so influential in directing the daily action on the weather map, migrates to the farthest south position in its cross-continental rush from the Pacific to the Atlantic. This permits polar airstreams from central and northern Canada to penetrate southward without hindrance and envelop a large part of the United States. Being relatively close to the source region of the cold air, the Green Mountain country receives the northerly airflow in an almost pristine state, with little modification of its frigidity.

During the days of January, the solar angle at noon increases its altitude above the horizon by 5°30', and the duration of daylight lengthens by about fifty minutes, resulting in an increase of 17 percent in the amount of insolation received over its annual nadir at the solstice in December, yet the increased thermal input does not make a sensible impression on the coldness of the Vermont atmosphere until the very last week of January. The "turn of winter" comes about January 26, more than a month after the solstice, with a gradual increase in the normal mean temperature following day by day. January 26 also marks the midway point of the winter heating season. Half of the winter's total of heating degree days should have been accumulated by then, and one should have at least half one's woodpile left.

In concert with the jet stream, the cross-continental storm tracks also attain their southernmost locations. The St. Lawrence Valley stormway continues to serve as the main path for cyclonic disturbances moving eastward from the Great Lakes. During their passage to the north, Ver-

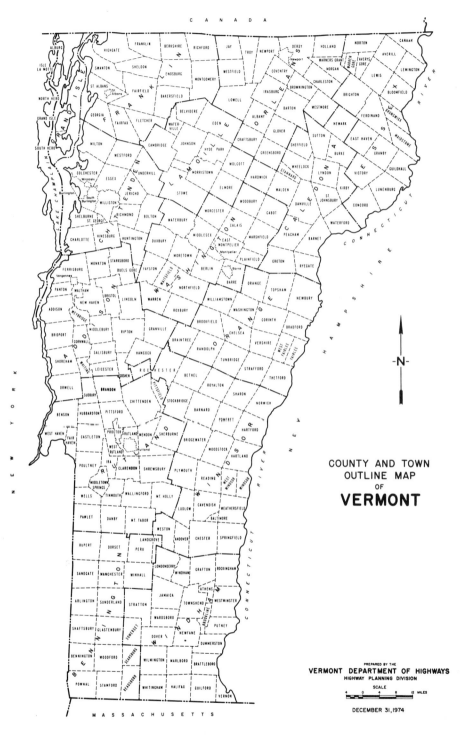

COUNTY AND TOWN
OUTLINE MAP
OF
VERMONT

PREPARED BY THE
VERMONT DEPARTMENT OF HIGHWAYS
HIGHWAY PLANNING DIVISION

SCALE

DECEMBER 31, 1974

THE VERMONT WEATHER BOOK

mont lies in a southerly airflow for a brief period. If the precipitation begins as snow, it may turn to rain for a while, then return to snow showers with the arrival of the cold front and northwest winds as the storm system to the north passes eastward.

Occasionally, storm centers coming from the Lower Lakes region or the Ohio Valley cross New England or pass just to the south in the vicinity of Long Island. Then a northeast flow prevails over the Green Mountains, and the precipitation falls as snow. At this time of year coastal storms (nor'easters or sou'easters) carry an abundant supply of oceanic moisture as they travel along the Atlantic seaboard and bring a threat of heavy snowfall to portions of Vermont. The exact trajectory of the storm center determines the distribution of snow over the Green Mountain region. In general, if the center passes inside Cape Cod over southeast Massachusetts, all of Vermont may get a heavy snowfall; if outside the Cape, only the southeast corner usually gets substantial amounts.

The principal high-pressure areas that direct atmospheric traffic across the continent have also retreated to southerly latitudes. The core of the Azores-Bermuda high-pressure area lies well offshore, allowing coastal storms to form along the South Atlantic coast and track northeast toward New England. The most frequented path of anticyclones over the interior of the continent from northwest Canada passes west of the Great Lakes. Vermont receives a strong northwest flow of very cold polar or of arctic air during each high pressure intrusion into the United States. Once the gusty winds subside and the atmosphere becomes relatively stable, sparkling days, clear and calm, ensue and supply a delightful interlude to the normally stormy midwinter scene. Sunshine during January increases to about 39 percent of the possible, a rise of 9 percent from the low point during the gloomy days of November.

Mean temperatures in January range from 13.8° at Newport in the northeast to 21.1° at Bennington in the southwest. The mean daily minimum at Newport averages the low figure of 3° and is probably close to zero or slightly below in the cold hollows that abound in surrounding valleys. All stations except Vernon in the southeast have a mean daily maximum below 32°. The highest January temperature reported during this century in Vermont was 70° at Dorset and Rutland in 1950, and the lowest, −44°, was reached at Enosburgh Falls in 1904 and 1914, and at Bloomfield in 1934. In the record coldest January 1970, on only two days, the 29th and 30th, did the thermometer rise above freezing in the northeast corner of the state at West Burke.

January receives much less precipitation than December, by 21 percent in the northeast and by 17 percent in the west and southeast sections. The southeast has more precipitation by about one third than the rest of the state, mainly because of its proximity to coastal storm tracks. Mountain stations, as usual, report the greatest amounts; locations above 2,000 feet in the south, exposed to an easterly flow of Atlantic Ocean

25

moisture, average as high as 4.5 inches, and valley exposures receive about 3 inches. Much drier conditions prevail across the north, where a continental airflow lacking precipitable moisture predominates; some station totals fall below 2 inches for the month.

The January thaw is an interesting feature in the midwinter calendar. Weathermen refer to its occurrence as a singularity comprising a reversal in the seasonal temperature trend. A thaw does not come every January, but statistical studies for the Northeast have demonstrated that warm spells may be expected toward the end of the month in a sufficient number of years to affect the normal trend of mean temperatures temporarily. A slight increase appears at the beginning of the last ten days in January and is most noticeable from January 23 to 26; then a slight decrease follows until a definite upward trend comes with the advent of February. Afternoon temperatures often rise into the forties and fifties during the January thaw, and nighttime readings may remain above freezing for a night or two. If the thaw is sustained, ice may break up in the rivers and lakes, creating ice jams and local flooding.

A typical January thaw occurred in 1972. On the morning of January 22, temperatures were generally below zero, though several locations registered single figures. The next day above-freezing conditions occurred at all stations, mainly in the high 30s with some western stations in the 40s. On the 24th, readings soared into the 40s, with a maximum of 50° at Chelsea. Thermometers again mounted to similar readings on the 25th throughout the state, but on the 26th only in the south were the 40s reached again. A cold front entering the northwest early that morning, pushed southeast during the day and overran the entire state by evening. During this period, temperatures over the state ranged from a low of −15° on the 22nd to 50° on the 24th. Readings were generally below zero again late on the 26th as the January thaw vanished.

Historic Weather Events in January

1 1918 Coldest New Year's Day so far: −24° at Northfield, with maximum of only −2°; below zero at Burlington for 94 consecutive hours, ending at noon.

   1970 Cold morning throughout state: −24° at Cornwall, −19° at Woodstock and Chelsea; −17° at Burlington and Rutland.

2 1870 Big south wind struck western sections; particularly destructive at Charlotte; many telegraph circuits knocked out.

3 1870 Winter freshet at Montpelier took out a temporary railroad bridge that had been erected after the great flood of October 1869 just three months earlier.

   1913 Vast cyclonic storm near Montreal set low-pressure records over the state: Burlington, 28.28″; Northfield, 28.35″.

4 1950 Warmest day in a very warm January: 70° in Rutland, 68° in Northfield, 67° in Middlesex, 63° in Burlington.

    1981 Bitter cold with temperature range at West Burke of −6° / −38°* on 4th, −15° / −38° on 5th; −18° / −35°, −1° / −37° at Mt. Mansfield; −19° / −31°, 4° / −20° at Newport.

5 1835 Coldest day of the "Cold Week" in a very cold winter: −40° at Montpelier and White River Junction; −38° at Bradford and West Burke; −26° at Burlington; −31° at Newport.

6 1959 High winds gusting from 40 to 80 mi/h drifted snow on highways; especially severe in the Northeast.

7 1968 Record early freeze-over of Lake Champlain at Burlington.

8 1841 High flood of Connecticut River at North Walpole, N.H., one foot above mark of September 1828; Hanover, N.H., measured deluge of 3.56" on 6-7th.

*Temperature readings expressed in this way give the maximum and minimum readings, respectively.

Using the town snowroller as a means of maintaining winter roads was the common practice well into this century. The hard packed snow made good sleighing and sledding possible. It usually required a four-horse team to pull the heavy snowrollers, and it was cold, hard work for both the team and the driver. Most farmers and rural residents didn't venture very far until after the snowroller had done its job.

9 1884 Shirkshire winds caused extensive damage in western valleys; church spires were toppled, barns and outbuildings demolished.

1968 Severely cold days: Bloomfield, $-10°/-35°$; West Burke, $-16°/-34°$; Mt. Mansfield, $-4°/-39°$.

10 1859 Coldest daylight hours so far recorded: Burlington's mean of $-27.4°$ was 4° colder than Jan. 23, 1857, and 8° colder than Feb. 5, 1855.

11 1859 Extreme cold continued: $-40°$ at St. Johnsbury and Randolph; $-39°$ at Woodstock; $-35°$ at Craftsbury; $-32°$ at Burlington; press reported $-55°$ at Island Pond.

12 1918 The only warm day in the month throughout the entire state during the severest winter of the century so far: at Enosburg Falls the temperature rose from 0° to 40°, then dropped to 0° again within 24 hours.

1968 At Burlington, 115 consecutive hours of below-zero temperatures ended with a rise from $-23°$ to 6°.

13 1914 Among the bitterest days of the century; at Northfield the temperature.rose from $-25°$ to only $-13°$; at Burlington, from $-23°$ to $-14°$; Bloomfield hit a low of $-44°$.

1964 Heavy snow and wind closed Route 9 from Brattleboro to Bennington; Readsboro had 21.3" of snow; Bennington, 15".

14 1934 Burlington's most concentrated snowfall: 24.2" in 24 hours.

15 1885 Wilson Alwyn Bentley of Jericho made his first microphotograph of a snowflake.

1957 Burlington's minimum of $-30°$ set a modern record (the maximum was only $-6°$); $-42°$ at Lemington; $-41°$ at Enosburg Falls; $-40°$ at West Burke.

16 1879 Snow fell from a cloudless sky at Newbury.

17 1820 Famous shirkshire windstorm did extensive damage in the valley of Bennington County.

1982 "Cold Sunday"; bitter northwest gale kept afternoon maximums near zero and drove minimums next morning to $-29°$ at Enosburg Falls and Chelsea.

18 1857 "The Cold Storm" raged over New England, with temperatures at zero or below; 10" to 12" of snow fell overnight; railroads and highways blocked by deep drifting snow.

19 1810 Famous Cold Friday in New England: at Hanover, N.H., a bitter northwest wind dropped the temperature to $-17°$ from 40° at noon on the day before; buildings were blown down; three children frozen at Sanbornton, N.H.

20 1961 Kennedy Inaugural Snowstorm: a coastal storm dropped 12" of snow on Vernon, but only a trace on Burlington to the north; elsewhere in the south falls of 8" to 15" raised snow depths to 20-30".

21 1887 Rapid rise of 81 degrees in 42 hours at Windsor: $-33°/48°$.

1957 January thaw after a long deep freeze; at Burlington, maximums of 50°, 60°, and 60° on the days, 21st, 22nd, and 23rd, respectively.

22 1976 Severe wind chill of −63° at Burlington; the maximum temperature, −5°, with 50 mi/h gusts.

23 1857 Cold Friday II: readings at Craftsbury were −34° in the morning, −23° at noon, −27° in the evening, with a mean of −28°; at Burlington, −30° minimum, with a mean of −23°.

1906 Record January thaw: 65° at Cornwall, 64° at Burlington and Wells River.

24 1857 Probably the coldest morning ever in Vermont: −50° at Montpelier and St. Johnsbury in press reports; −44° at Norwich; −54° reported in the upper Connecticut River Valley near Stratford, N.H.

25 1888 First Blizzard of '88 began: a deep storm center moving east over Massachusetts caused northwest gales and heavy snow; Newport measured 22"; Woodstock, 16"; Burlington, 12". The Rutland Railroad was blocked near Charlotte for four days by massive drifts from 15 to 20 feet deep.

Contrasting sharply with winter road maintenance in the days of the snowroller, today's huge wing plows keep the highways open even in the worst winter weather. Thanks to state highway crews and excellent equipment, Vermont has a reputation for year-round open roads.

26 1839 Great Southeasterly Storm and Thaw of 1839: a deep cyclonic storm over New York State caused a surge of southerly air that raised temperatures from below zero to the 50s; massive ice breakup caused high floods, with extensive damage to bridges, dams, mills, highways.

27 1927 Massive anticyclone with barometer readings of 31.03" at Burlington, 31.00" at Northfield, and temperatures of −27° at St. Johnsbury, −22° at Northfield, −20° at Burlington.

28 1844 At Northfield, −42° minimum during the coldest week of the first half of the nineteenth century.

1925 Coldest day of a very cold month: −42° at Enosburg Falls, −39° at Garfield*, −30° at St. Johnsbury, −26° at Burlington.

29 1973 Heavy snowstorm in the southeast: Vernon had 15"; Readsboro, 14.3".

30 1925 Big snowstorm of an especially wintry month, Burlington had 15.1", and a total of 31" for the month and a temperature departure from the normal of −6.8 degrees.

31 1857 Blizzard conditions at Huntington described by James Johns as "a real devil's training day" in his unique diary, which treated many weather matters with native humor.

1888 Snow accumulation at Northfield reached 54" depth after a very stormy month.

February

"Let it snow, let it snow, let it snow." The words of the popular song of the 1940s provide the weather theme for February in the Green Mountains. Since nothing much can be done about it, one should relax and enjoy the beauties of the white countryside. It is a short month, and soon the Snow Kingdom will be in retreat.

Vermonters' attitudes toward snow have changed over the years. In the pioneer days, they regarded snow as a great benefit since it enabled sleighs to glide with ease from place to place. Heavy hauling could be accomplished on the farm in much less time than when the ground was bare and wheels were necessary. With the advent of the automobile, snow became a nuisance since it had to be plowed off the roads and driveways at a great expenditure of time and money. In recent years snow has again returned to high esteem. Out-of-state skiers throng the Vermont hills and bring millions of dollars in revenue to an otherwise slack business season. It is with truth called "white gold."

Though having fewer days by two or three, more inches of snow ac-

*Garfield is a village in the town of Hyde Park.

tually fall in February than in any other month, and the percentage of the total precipitation falling as snow reaches the highest figure of the year: 97 percent on the first day of the month, and 88 percent on the last day, according to a study of snowfall at Burlington.

The sun makes good progress northward during February. It climbs about nine degrees toward the zenith until on the 28th it stands at noontime about 37 degrees above the southern horizon. The period of direct sunlight is 71 minutes longer at the end than at the beginning of the month, and total insolation increases by 12.8 percent.

February is the stormiest month of the year in Vermont. The main jet stream crosses the Pacific Ocean along latitude 35°N; when approaching the continent, it dips southeast over Baja California. After taking a southerly route across the country, the jet trends northeast to the Ohio Valley and is joined by a weaker westerly jet from the northern plains. The combined airflow leaves the continent in the vicinity of the Virginia capes. The southerly displacement of the cross-continent air movement contributes to storm formation (cyclogenesis) over the tropical Gulf of Mexico and Texas. During the winter season, storm centers from this region head northeast along either the western or eastern slopes of the Appalachian Mountains.

The path of the storm track determines what kind of weather will prevail

Skating, tobogganing and ice yachting on Burlington Bay, Lake Champlain. From the Burlington Daily Free Press, Carnival Number, *1887.*

31

over Vermont. If the storm centers pass to the west over Pennsylvania and western New York State, south and southeast airstreams will overspread Vermont with a steadily increasing warmth. Though the storm precipitation may begin as snow, the possibility of a change to ice pellets and eventually to rain increases with the continuation of the flow from the southerly quadrant of the compass. If the storm track stays to the east along the Atlantic coastal plain, precipitation will begin as snow and usually continue in that form, perhaps becoming heavy when the moisture of the nearby ocean is tapped by the circulation.

The major stormway from the Great Lakes down to the St. Lawrence Valley remains active, though less frequented than in the earlier winter months. Cyclonic systems moving to the north of Vermont come from relatively dry regions and are usually fast movers, so the period of precipitation is shorter and less intense than with Atlantic coastal storms.

The behavior of polar anticyclones continues the same as in January. The frozen surface of Hudson Bay and the snow-covered tundra of central Canada condition the atmosphere for several days until the air masses move into the northeastern states on gusty winds. Occasionally, high-pressure areas of great magnitude build up in the Hudson Bay region. Known as Hudson Bay Highs, they await the passage of a cyclonic disturbance along the Atlantic seaboard before their frigid airstreams are enticed southeast into the rear of the departing storm system. They treat the Vermont countryside to its coldest weather, with the temperature dropping as much as 30 to 40 degrees to minimums of $-20°$ or lower. The record cold of February 1934 resulted from a series of cold outbreaks from the Hudson Bay region following the passage of successive snowstorms along the coast.

In February, mean temperatures begin to respond to the increasing altitude of the sun. Newport in the northeast shows a rise of 2.3 degrees to 16.1°; Vernon in the southeast increases by 2.3 degrees to 22.5°. The highest readings obtain in the southwest: Bennington reaches a mean of 24.1°. The severest cold can occur as late as February 16, but thereafter comes a decline in the refrigerating ability of the Hudson Bay region. Temperatures in Vermont do not drop below $-40°$ after mid-February. The worst of winter's sting is over.

Total precipitation in the northeast and southeast sections averages slightly more each day than in January because of more plentiful Atlantic moisture. The west and northwest sections are the driest under the continued domination of the westerly airflow. Burlington receives only 1.68 inches in February, while Vernon has an expectancy of 3.09 inches. The mountain stations catch the maximum amounts; Searsburg Station, at a high elevation in Bennington County, averages 3.73 inches.

The highest February temperature reported in this century was 65° at Rutland in 1957 and the lowest, $-46°$, at East Barnet in 1943.

The second day of February marks a big event on the folklore calen-

THE VERMONT WEATHER BOOK

dar, a day thought to portend a possible end of wintry conditions since the Europeans regarded November, December, and January as the three winter months. Hibernating bears and badgers were supposed to come out of their lairs and check the weather prognostics. In America the role was transferred to the groundhog, or woodchuck, of which Vermont has an abundance. If it is sunny on February 2, the animal will see his shadow and be frightened into returning to his hibernation hole for six weeks longer, and winter will continue for that period. Since the percentage of possible sunshine is 46 percent at Burlington in February, the groundhog has about a fifty-fifty chance of seeing his shadow.

Historic Weather Events in February

1 1920 Massive anticyclone set a high barometer record for the state: 31.14″ at midnight of 31st-1st at Northfield; Burlington reached 31.12″. Temperatures: −39° at Enosburg Falls, −38° at St. Johnsbury, −28° at Burlington.

2 1920 Temperature surge from −39° on the 1st to a maximum of 47° on the 2nd at Enosburg Falls; from −33° to 48° at Northfield.

1962 Coldest in 30 years over the north: −41° at Enosburg Falls, −40° at Bloomfield, −25° at Burlington, −21° at Dorset; too cold for groundhog watching.

1973 Glaze storm throughout state: 1.34″ at Ludlow, 2.22″ at Searsburg.

3 1961 Climax of a prolonged cold spell that began on Jan. 18; continuously below zero for 48 hours on 1st-2nd at Burlington; Newport readings, −3°/−21°, −1°/−19°.

4 1972 "Severe storm of near hurricane proportions" swept the state; 60 mi/h gusts at Bennington and St. Johnsbury; 0° at Montpelier and Newport.

5 1886 Cold day: −40° at East Berkshire, −30° at Newport.

6 1855 Cold Tuesday: −44° at West Randolph; Hanover, N.H., had its coldest day since 1835 with a mean of −19.4°.

1920 Great storm of snow and ice pellets, which began on the 5th, dropped icy covering with the temperature in the 20s throughout the state; Somerset reported 4.20″ melted precipitation, Cavendish, 2.42″, Woodstock, 2″.

7 1805 Both arches of the brand-new Brattleboro bridge over the Connecticut River collapsed from the weight of the deep snow.

8 1861 One of the coldest daytimes in Vermont history: Lunenburg read −40° at sunrise, −22° at 2 P.M., −30° at 9 P.M., with a mean of −30.6°; Burlington's minimum of −32° was coldest minimum ever known.

Skiing was a new and somewhat daring sport when Florence Clark set out to give it a try on the Camp Comfort golf course in Brattleboro in 1915. With ski tows yet to be thought of and skis much longer than today's trim models, the familiar snowshoes offered the best way to climb hills.

9 1918 Long cold spell ended; Burlington experienced minimum of 10° or lower every day from January 9 through February 9.

   1934 Coldest morning statewide in the twentieth century: −41° at Bloomfield and East Barnet, −39° at Enosburg Falls, −36° at Woodstock, −30° at Vernon, −26° at Burlington; temperatures remained below zero all day.

10 1857 Devastating freshets in early breakup of winter when very warm February followed very cold January.

11 1780 After nine weeks of continuously cold weather, a thaw arrived; eaves at Barnard began to drip for the first time (January 1780 was probably the coldest month in New England's thermometric record).

12 1899 Coldest day in a spectacularly cold week: Woodstock, −2° / −30° (below zero every morning, 9th to 16th); Bennington had six mornings, 0° or below.

13 1899 The Cold Storm of '99: Woodstock had 15″ of snow with temperature ranging from 0° to 6°; Derby in north had 12″; Vernon in south, 14″.

   1959 Blizzard conditions in Franklin County, "worst in years"; temperature ranged near zero; flow of milk from farms to cities delayed.

14 1888 Spectacular "sudden change" at Northfield: drop of 49 degrees to −29° in 40 hours, followed by rise of 68 degrees in 32 hours on 15th to 17th.

THE VERMONT WEATHER BOOK

15 1885 High winds accompanied statewide heavy snowstorm; railroads blocked; 18" at Burlington; 48" on ground at Chester.

16 1943 One of the coldest mornings of the century: East Barnet −46°, Bloomfield −45°, Cavendish −40°, St. Johnsbury −43°, Hanover, N.H., −40°.

17 1896 Severely cold day, below zero all daylight except in southeast; Woodstock −30° on 17th, −36° on 18th.

1958 Greatest snowstorm of the mid-century: Hanover, N.H., 31.3", Readsboro 25", Cavendish 24", Montpelier 20", Burlington 14.9".

18 1952 Big snowstorm: 28" fell at Northfield, making 40" depth; 23.5" at Somerset, 48" depth; and 22" at Bloomfield, 40" depth.

19 1972 Severe blizzard raged over all of state: Readsboro 17.2", St. Albans Bay 15", Vernon 12".

20 1934 Blizzard conditions prevailed, worst storm of the coldest month in modern records; temperature ranged from 11° to −2° at Burlington; 14" snow at Cavendish, 10" at Rutland.

21 1979 String of eighteen consecutive days with below-zero minimums ended at West Burke; average minimum −24°; lowest −37° on 12th and 13th.

22 1958 Climax of long below-freezing period: Newport not above 32° from February 1 to 23, West Burke from February 3 to 22.

23 1972 Late cold: Enosburg Falls −34°, West Burke −34°, Newport −30°.

24 1936 Brown snow fell in north, attributed to duststorms on Great Plains.

25 1866 High flood on Connecticut River; bridge at Windsor carried downstream, hit Sullivan railroad bridge and carried it along.

26 1957 End-of-February thaw closed hard winter; thermometer soared to 68° at Bennington, 65° at Rutland, 60° at Burlington.

1981 False spring: temperature continuously above freezing at Burlington from 16th through 26th; maximum 62° on 19th, above 58°, 18th to 22nd.

27 1829 Fayetteville, now Newfane: heavy snow of 30" in two storms, made 60" on ground, according to General Martin Field's meteorological record.

1963 Late-season cold: −30° at Enosburg Falls and West Burke, −29° at Bloomfield, −25° at Woodstock, −21° at Burlington, −33° at Colebrook, N.H., on upper Connecticut River.

28 1893 Record snowy month in south: 53" fell at Monroe on Massachusetts border near Readsboro; similar to great snowfall in early March 1947.

29 1832 Snow depth of 36" at Burlington; according to Professor Zadock Thompson's weather notes, exceeded the modern record of 33" set in 1969.

THE VERMONT WEATHER BOOK

1984 marked the fiftieth anniversary of the nation's first ski tow on Gilbert's Hill just north of Woodstock, Vermont. A simple rope tow, powered by an automobile engine, it attracted hundreds of people who had become acquainted with downhill skiing as a sport but disliked the long climbs between runs. Wallace "Bunny" Bertram is credited with making the tow a success. He later moved the tow to a South Pomfret hill now known as "Suicide Six." From this simple beginning a multi-million dollar recreation industry evolved that has changed Vermont's economy, appearance, and way of life. Photos courtesy Woodstock Inn and Resort.

March

"Lion-like March comes in hoarse, with tempestuous breath." So William Dean Howells described the blustery month. March is aptly named after the Roman god of war, for its windy activity results from a combat for possession of the New England region between the retiring forces of the North and the advancing forces of the South. "Like an army defeated, the snow hath retreated," was William Wordsworth's observation of the usual March weatherama.

The front of the March battle zone often lies right over Vermont, creating much storminess and causing rapid changes from cold to warm and back

37

to cold again. Temperature contrasts as great as 20 to 30 degrees may exist across the weather front, creating distinct thermal and precipitation zones. Rain may be falling on one side of the mountains while snowflakes are descending on the other side. Or the battleline may migrate north, then south, giving a locality periods of frozen or melted precipitation alternately. Often the combat zone has a vertical alignment: higher ground may receive a white covering while valleys experience a cold rain.

The main event in March is the beginning of astronomical spring at the vernal equinox, which falls on or about March 21. Having moved north to the equator, the sun rises directly in the east and sets in the west; hence night and day are about equal. The term *equinox* is derived from the Latin *aequi* (equal) and *nox* (night) (the words might just as well be *aequi dies*, or equal days). That the equinox has little to do with the daily weather led Dr. Henry Van Dyke to compose a warning about the spring climate: "The first day of spring is one thing, and the first spring

For years it was a Vermont truism that in addition to the normal four seasons the state had a fifth - mud season. It still exists, but thanks to the prevalence of "hard" roads and improved methods of building and caring for gravel roads, mud season is simply an occasional nuisance to the driver and not a serious threat to travel.

38

day another. The difference between them is sometimes as great as a month."

The principal jet stream in March enters the continent over Oregon and dips southeast to Missouri before heading directly east to follow the eastern part of the February track over Virginia. The southern jet, of such importance for storm making in January and February, gradually loses strength with the approach of spring and by the end of March has become a minor influence.

March is a transitional month, when storm tracks retain many of winter's aspects yet at the same time start to exhibit some of the characteristics of spring. During March storms tend to form more often over the continent and less often over the ocean waters. Both principal storm tracks of February—down the St. Lawrence stormway and along the Atlantic seaboard—remain active, although they develop less frequently.

The daily progress of coastal disturbances along the Atlantic seaboard is influenced in March by the presence of blocking anticyclones over Quebec and the Atlantic provinces of Canada; their position slows the northeast movement of storm centers while at the same time increases their intensity and lifespan. Examples of severe storminess in March abound. Often the "Crown of Winter Storm" comes in early March and raises snow depths to the season's high. The Great Blizzard of '88 is the supreme example of a stalled coastal storm keeping the precipitation process going about double the number of hours that an unfettered storm can and in so doing producing about twice the amount of snow.

Although the principal anticyclonic tracks during March retain most of the characteristics of the winter months, several typical spring features are introduced. Of importance to the Green Mountain region is the tendency of Canadian high-pressure areas to develop farther east, over central Canada, and to track east of the Great Lakes rather than west. The Hudson Bay High is often the parent of the blocking high-pressure areas over eastern Canada, mentioned above as having important influences on the duration and intensity of New England coastal storms.

With the increased elevation of the sun and the longer extent of daylight, mean temperatures take an upward jump of about 10 degrees during March. In the northeast, Newport averages 27.1°; Burlington, in the northwest, 29.2°; and both Vernon, in the southeast, and Bennington, in the southwest, just above 32°. Heat waves in March have driven the mercury as high as 84° at Bennington in 1945 and at Burlington in 1946, and cold waves have dropped readings as low as −37° at St. Albans in 1938.

March brings an increase in precipitation over the relative dryness of midwinter. Several mountain stations in the south average over four inches, topped by Searsburg with 4.69 inches. In the north increases are of smaller degree; some stations measure about two inches—Gilman 2.05

39

inches and Burlington 1.93 inches. Most stations at middle and low elevations report almost three inches. March can be a very wet month as the statewide average of 6.92 inches in 1980 demonstrated.

March brings the high rite of the Vermont spring—sugaring time. The metabolism of a maple tree is a most complex matter, and the stages of development are intimately tied to the prevailing weather conditions. Warm days in the 40s and cold nights in the 20s are basic requirements, and other considerations such as humidity, precipitation, cloud cover, wind direction, and solar intensity play a part. The annual spring contest between frost and sun is crucial. Scott Nearing, the author of many works on country living, first learned about the intricacies of sugaring from his own Vermont grove. His first boilings on the day of the initial tree tapping took place on dates ranging from February 23 to March 25 with the mean date falling on March 12-13. The length of the sugaring season averaged forty-one days. More sugar was made in April than in either March or May. The quality of the sap deteriorated rapidly in the absence of overnight freezes, and the run ceased altogether with the leafing of the trees. Increasing warmth was responsible for the commencement of the sap flow, and its continued increase also caused the ending of the sap run.

Historic Weather Events in March

1 1804 An excessive snow load crushed the new bridge over the Connecticut River at Norwich.

2 1900 The two-day Great March Storm of '00 ends—the "worst since March 1888". At Northfield, 30" of new snow in 36 hours; at Woodstock and Derby, 40"; at Burlington, 14".

3 1947 The deepest snowstorm of modern times: Readsboro had 50" from 2nd to 4th and a record 23" in 24 hours; 20" at Whitingham farther north; rain fell in valleys.

4 1938 Record March cold: −37° at St. Albans, a New England record low for March; below zero three mornings; −32° at Bloomfield, Enosburg Falls, and Newport; at Burlington, −24°, a late-season record.

1950 Cold day: −34° at West Burke; −32° at Chelsea and East Barnet; Enosburg Falls below zero on first seven mornings of month.

5 1872 Severely cold day: Lunenburg had −25°, −10°, and −18° at the three observation hours: "5 & 6 were the coldest days of the winter. The wind at so low a temperature made it impossible for people to move about much of the day," Hiram Cutting in his record book.

6 1920 "Worst storm since the March blizzard of 1888 is passing into history," *Free Press*; at Burlington, 14" of new snow, making 24" depth, both March records; highways blocked; Barre isolated; no daily mail at Brandon – first time in history.

1959 Straightline wind rush from Underhill Center to Cambridge; covered area 10 miles long, 2 miles wide; barns unroofed; TV antennas downed.

7 1971 Second heavy snowfall in a week left a total of 88" on the ground at Orange; 19" at Montpelier on 7-8th, after 17" on 3-5th, raised depth to 70" at airport.

8 1967 57" of snow on ground at Readsboro near Massachusetts border.

9 1961 Snowstorm of 6-20" caused numerous accidents; at Peru, 12"; at Readsboro, 26" on ground.

10 1964 Snowstorm of 8-20" closed airports, hampered highway traffic; at Newport, 18.6" on 9-10th; at Burlington, 12.4".

11 1936 First rains leading to the Great March Flood of 1936 fell on snow cover with a water content up to 16". At Cavendish, 3.25" rain fell in 24 hours; at Brattleboro, 2.15"; at Woodstock, 2.03".

12 1888 March Blizzard of '88, Vermont's greatest snowstorm: began about noon on the 11th, and continued for about 48 hours. In northeast, 18" at Newport; in the northwest, 23" at Burlington; in the southeast, 38" at Vernon; in the southwest, 48" at Bennington; gales from north and northeast on 12th and early 13th; lowest temperature, 4°, in southwest at Manchester; warmest in northeast at Newport, 19° on 12-13th.

13 1936 Snow rollers observed at Burlington; the muff-like cylinders varied in size up to 13" diameter; temperature was at 32° and wind at 27 mi/h.

1962 Wet snowstorm of 5" to 18" over the state caused great tree damage and power outages. Shirkshire winds in Bennington Valley hit 81 mi/h in gusts; a brick building was demolished; a car was blown from the road near Rutland.

14 1850 A thunderstorm at Newbury with deep snow on the ground; several buildings were struck by lightning.

1984 Very deep snowstorm in mid-March set all-time 24-hour record at Peru where 37" was measured on 14th; at Rochester, 30", at Cavendish, 24", at Rutland, 23"; at Chelsea, Manchester, and Union Village Dam, 22".

15 1937 Heavy snowstorm across the state after a very mild winter; Burlington received 19.5" of snow on 15-16th, with temperatures 29°/20° and 33°/27°.

16 1862 Major storm of snowy winter: Burlington 20", Craftsbury 19.5", Lunenburg 17"; storm built up very deep snow cover with high flood potential; thick crust preserved great depth until mid-April.

Spring thaws bring on mud time. Here Clinton Gilbert of Woodstock drives his team down the oozing road toward his barn in the 1940 photograph from the F.S.A. collection in the Library of Congress.

      1983 Record deep late-season snowstorm on 16-17th; Burlington had 15.6" total, Mt. Mansfield 16", Waterbury 14.4".

17 1956 Snowy March's biggest storm in 1956: 21" at Northfield raised depth to 60", Mt. Mansfield to 58", Somerset to 52".

18 1961 Pre-equinox cold wave: −14° at West Burke, −13° on Mt. Mansfield; −7° at Burlington, 2° at Vernon and Bellows Falls.

19 1936 More than 3" of rain caused second rise within a week and greatest flood ever known on central and southern Connecticut River; record high stage at White River Junction; top of Vernon dam submerged under 11' of water.

20 1903 Burlington thermometer reached 70°, earliest seasonal reading in the seventies.

      1971 Third heavy snowstorm of month: 19.8" at Cavendish raised depth to 45"; Readsboro's 15" produced 70" depth on ground.

21 1801 "Very copious and severe rains" raised Connecticut River at Walpole, N.H., higher than known for 20 years; called Jefferson Flood by Federalists who blamed the disaster on the new Democratic president.

22 1955 Small tornado at Shaftsbury: two barns and a silo leveled; wires down; path 65 yds long, 33 yds wide.

23 1934 Late cold: −8° at Newport, −4° at St. Johnsbury and Woodstock; 1° at Burlington.

24 1826 Freshet on Connecticut River, also on West, Williams, and Saxtons rivers; all bridges between Newfane and Brattleboro out; highway between Montpelier and Randolph washed out.

25 1969 Heavy wind storm, rain in valleys, snow on high elevations on 25-26th; Cavendish 2.19" precipitation; Peru 9" new snow; only a trace at Burlington.

26 1847 Greatest snowstorm of mid-century: "30 inches by actual measurement" at Barnard; 17" at Burlington in "driving storm"; roads badly drifted, blocked for two days; barometer at 28.73" on 27th; 22" snow at Lunenburg, which remained throughout April.

27 1913 Highest flood ever known at Brattleboro; three million feet of logs carried downstream on White River, smashing railroad bridge; Woodstock reported 3.72" rain, St. Johnsbury, 4.40".

28 1919 Heavy snowstorm: Burlington 19.1" on 27-29th equaled 1.89" melted precipitation; temperature dropped from 62° to 15°; Northfield measured 3.62", Somerset 3.32", Woodstock 2.94", Cavendish 2.46" in 48 hours.

29 1946 Burlington soared to 84°, first seasonal reading in the eighties; warm second half of month: Bennington maximum above 58° from 13th to 30th; above 60° from 18th to 30th; above 70° from 26th to 29th; range on 29th from 34° to 84°.

30 1951 Burlington clocked 56 mi/h sustained wind from south for a March record.

31 1886 Strong south wind caused ice breakup and snow melt; Winooski River flooded streets of Montpelier and Berlin; railroad bridge carried out; high floods on Lamoille and northern Connecticut tributaries; railroad lines cut.

April

"Oh the lovely fickleness of an April day." The words of W. H. Gibson well express the contrariness of the weather elements in the first full month of spring. April often opens in the cold lap of a late winter and ends in the warm embrace of an early summer. Mud may be everywhere.

The sun accelerates its upward climb this month almost to a sprint, increasing by 10°15′, so that at noon over central Vermont on the fifteenth the sun stands about 54°30′ above the horizon. The day lengthens by about 120 minutes, adding an hour more of daylight at each end.

During April, with the axis of the main westerly jet stream remaining quite far south at 38°N, the storminess of late winter continues. An area of high cyclonic frequency lies off the Middle Atlantic coast, and some of its disturbances affect the Green Mountain area. The main storm tracks

43

of April resemble those of March: the St. Lawrence Valley path has frequent travelers whose frontal systems trail south over New England and bring quick changes from warm to cold to warm, and from dry to wet to dry.

The Hudson Bay High is now the dominant feature of weather maps over eastern Canada; its high-pressure ridge, which extends over the ocean, continues to block the movement of storm systems from the west and along the Atlantic coastline. This pressure alignment often feeds cold airstreams from Labrador and Newfoundland southwest into New England and is directly responsible for the periods of backward spring weather so characteristic of the region.

A center of maximum anticyclonic activity in April is situated over South Dakota; airstreams from this source that arrive in Vermont are modified by their trip over bare ground and do not have the sharp crispness of the air flow from the highlands east of Hudson Bay.

The thermal onrush of the new spring season outdoes the increase in solar elevation. Mean temperatures rise from 12 to 14 degrees in various parts of Vermont during April. At Burlington, for instance, the mean

Spring mud season as well as general hard use made it difficult to keep Vermont's early dirt roads in good condition. Many towns owned and used their own road scrapers to help restore the roads. Initially pulled by single or double teams of horses as shown in this picture taken in Jericho, Vermont, the machine evolved into today's powerful, motorized road grader still used in many Vermont towns. From the W. S. Bentley collection, Jericho Historical Society.

THE VERMONT WEATHER BOOK

maximum rises from 46° on April 1 to 60° on April 30, and the mean minimum climbs from 27° to 38° with a freezing night being an exception by the month's end. Newport in the northeast has a mean of 40.2°, while Vernon in the southeast is 44.9°, and Bennington in the southwest, 45.8°. In April temperatures have ranged from a maximum of 97° at Vernon in 1976 down to −12° at Bloomfield in 1923.

A mean temperature of 45° is the threshold of growth for much of nature's plantlife. A harbinger of spring is the appearance of pussy willows, which require a few days with warmth in excess of 45° to get started. The "pussy" is not a flower, but a bud, said to resemble a number of small kittens climbing the stem. Though pussy willows may be brought indoors and honored as a decoration in April, they must give way in May to more colorful productions of nature.

The precipitation pattern begins to change from a winter to a spring type in April. By the beginning of the second week at Burlington, the chance of snow drops below 50 percent and by the end of the month well below 25 percent. Convective-type showers and thunderstorms increase, which swell the precipitation totals well above the winter amounts. The mountain stations continue the wettest with average amounts about 4.50 inches, and valley stations show substantial increases: Vernon to 3.84 inches and Bennington to 3.56 inches.

"April showers that bring May flowers" play an important role in nature's calendar. After the general storms of winter with steady precipitation over wide areas, the rainfall pattern becomes more localized, with brief showers over limited areas predominating. The increase in the sun's heat causes the surface air to warm, while the air aloft remains relatively cold. This sets in motion the process of convection whereby a parcel of warm air rises in the same manner that a hot air balloon does. In rising, the air cools at a steady rate until it reaches its saturation temperature, or dew point. Its moisture condenses into a visible cloud. If conditions continue favorable, the cloud will grow and turn into a cumulus shower cloud or even a cumulo-nimbus thunderhead. April is usually the first time in the season that sufficient warmth is available in the atmosphere to start the local cloud-making process, which is why showers and April have become synonomous.

Historic Weather Events in April

1 1807 Great April Fools' Day Storm. "Snow! Snow! Snow!" was the lead in the Randolph *Weekly Wanderer;* editor could not publish for two weeks due to lack of arrival of news; Danville received 30", making a depth of 60"; Marshfield accumulated a depth of 54"; Lancaster, N.H, 45"; Montpelier, "near four feet."

1923 Record April cold: Bloomfield −12°, Garfield −8°, Bennington 4°, Burlington 5°.

2 1975 Vast storm system battered all New England for three days with gales and heavy snow; extensive damage to trees in Vermont; Enosburg Falls 24.4", Peru 30", Manchester 21" new snowfall.

3 1977 Wind gusts from 40 to 70 mi/h downed trees and wires; extensive damage to Burlington piers along lake.

4 1899 Sleighing at Hanover, N.H., became poor after a record 129 days of good snow surface; 40" fell in March to boost season total to 96".

5 1954 Late-season cold: −17° First Conn. Lake, N.H.; −12° at Lemington; −10° at West Burke; −9° at Bloomfield on 4th or 5th.

1963 Wind gusts to 65 mi/h over state; widespread damage to trees; Burlington hit 42 mi/h from northwest.

6 1982 Spring blizzard dropped 25" on southern section; gales followed for two days causing severe drifting and blocking highways. Light snow across north; Burlington only 0.1".

7 1873 Dirty snow fell having a black appearance; attributed to Great Plains dust storm carried east at high altitude.

1972 Late-season cold: −2° at Enosburg Falls, 0° at Newport and Northfield. Burlington's 2° was April and late-season record; snow still 39" deep at Peru.

8 1929 Record early heat: Vernon 90°, Brattleboro 89°, Bennington 85°, Woodstock 83°, Burlington 71°.

9 1956 Snowstorm in south: Somerset 11", making depth 47"; only trace at Burlington and Newport.

10 1945 Early heat wave: Burlington 76°, 81°, 84°, 82°, 84°, from 9th to 13th.

11 1948 Windy day: Burlington had 18.9 mi/h average with peak 52 mi/h from the south.

12 1929 Spring snowstorm: 14" Bloomfield on 12-13th; Burlington 5.5"; Somerset had April total of 35".

13 1933 Heavy snowfall on high ground; 20" Cavendish and Woodstock; only trace St. Albans, 0.8" Burlington.

1945 Early heat wave: 89° Bennington, 80° or above from 10th to 14th; Bloomfield in Northeast Kingdom, 85° on 12th and 13th.

14 1962 Heavy wet snow, measuring from 12 to 18" in mountains, damaged utility lines.

15 1895 Heavy rains of 50 hours duration caused greatest flood since 1869 on Winooski and Connecticut rivers; Brattleboro had worst inundation since 1862; log booms broke, bridges damaged.

1972 Heavy wet snow caused damage in triangle from Brattleboro and Bellows Falls over the mountains to Bennington, "the worst utility damage since 1938"; Peru 6.5", Readsboro and Vernon 3" snowfall.

16 1852 Late-season snowstorm left 18" at Newbury, 16" at Randolph; 6" at Burlington; Hanover, N.H., had 33" total in April.

17 1843 After protracted winter, heavy rain fell on deep snow cover: Hanover, N.H., 1.37" on 14-17th with temperature in the 40s; great flood followed.

1983 Record late-season snowfall set April record at Burlington with 15.6"; Mt. Mansfield had 16", raising depth on ground to 81".

18 1967 Thunderstorm occurred during snow and ice storm, part of five-day wet period; Peru 7"; Somerset 33°/30° on 18th.

19 1862 Great snowmelt flood after winter of deep snow cover; ice already out and very little rain fell; Connecticut rose 5.5 feet.

1976 Unprecedented April heat; Vernon 92°, 97°, 96° on 18th to 20th; Burlington hit 91°, Montpelier 90°, Mt. Mansfield 72°.

20 1953 Late-season heavy snow on higher elevations: Somerset 15.5", Enosburg Falls 13.5", Burlington 4.6", Bennington trace.

21 1966 Gusts to 50 mi/h did extensive damage, unroofed church at Winooski.

22 1814 Ice break-up at Montpelier carried out highway bridge to Berlin.

23 1869 Great spring flood: Otter Creek rose to record crest at Brandon after 1.76" rain caused snow melt; Connecticut River highest ever at Bellows Falls except in 1862; railroad tunnel dammed to prevent street flooding.

24 1852 "Greatest flood ever known" at Wells River after sudden melting of snow; Hanover, N.H., reported 4.87" of rain in April.

1913 Remarkable April heat wave: 92° at Cavendish, 88° at Hanover, N.H., 85° at St. Johnsbury on 25th.

25 1866 High spring flood on Passumpsic River in Caledonia County destroyed four bridges, washed out dam, and closed five factories at St. Johnsbury.

26 1874 Heaviest snowstorm of snowy April; Bellows Falls 24", Chester 18", Woodstock 15", Bethel 15", Ludlow 14", Burlington 9".

27 1850 Flood at Wells River was highest since April 1807; at Newbury, where all intervale oxbow covered except 2 to 3 acres, since October 1785.

28 1962 Early-season heat wave: Bennington 89°, Burlington 87°, breaking existing records.

29 1854 Heavy rains set stage for great flood; St. Johnsbury 2.60" on 26-30th with 91° on 30th; Hanover, N.H., 3.47" on 28th-1st of May; Connecticut River reached the highest of record to that date in Massachusetts and Connecticut.

30 1874 Woodstock received another snowstorm of 10" as final installment to record April total of 49". "It is not every April 30th that sleighs are running in the streets of Woodstock" (*Vermont Standard*). St. Johnsbury reported "nearly three feet" on ground.

May

In view of its northern clime, most years in Vermont would be more in tune with nature if May Day was celebrated at the end of the month when the warming of the sun has actually produced a variety of flowers. In May the northward move of the direct rays of the overhead sun has progressed into the Caribbean Sea area, an advance of almost seven degrees during the month. The period of daylight between the rising and setting of the sun lengthens by another hour; by the end of the month we enjoy over fifteen hours of sunlight.

The path of the undulating jet stream across the eastern United States reflects the changing season. The increasing influence of the Bermuda-Azores high-pressure area becomes evident with the northward shift of the jet stream's location, whose central core now flows northeast from the Ohio Valley, over southern Pennsylvania, to the vicinity of Long Island south of New England. However, this is the mean course of the jet stream, which actually shifts its position north and south many times as it undulates eastward, from day to day and from week to week.

The speed of cyclonic disturbances crossing the United States and Canada slows and their intensity diminishes in May. The path from the interior of the continent now traverses Ontario and southern Quebec to the north of the St. Lawrence Valley, though their fronts still extend southward over Vermont and bring frequent weather changes. The track along the Atlantic seaboard practically disappears as a weather-making factor for interior New England.

The Bermuda-Azores high grows larger and stronger. Its lateral axis now lies along 40°N latitude and the north-south dimensions expand. The influence of the Hudson Bay High gradually lessens as the month progresses, but occasional blocking of the normal movement of storm centers still remains a factor to be reckoned by the forecaster.

The isotherms, or lines of equal temperature, continue their northward progress as in April. All of Vermont except the high elevations have means above 50°, ranging from 53.1° at Newport in the northeast to 56.5° at Bennington, a smaller difference between north and south than in the winter months. During May the thermometer has ranged up to 95° at Bloomfield and Cornwall in 1929 and Bellows Falls in 1962, and down to 14° at Bloomfield in 1946. The anomaly of having the highest temperature in the state registered at the most northerly station may be explained by the fact that air in May moves over heated land, and the station with the longest overland trajectory for southwest winds therefore will have the highest temperatures.

All sections of the state show increases in precipitation over April: the northeast by 17 percent, the west by 13 percent, and the southeast by 6 percent. The greatest amounts fall in the mountains where monthly

Dewdrops sparkle along the edges of a wild strawberry plant in this Bentley photograph. Note the carefully arranged background. From the W. S. Bentley collection, Jericho Historical Society.

catches range up to 4.53 inches at Searsburg Station in Bennington County. Totals are less in the north: Newport has 3.22 inches and Burlington 2.96 inches. Most of the increases result from more frequent thunderstorm activity; three are expected in the month of May in the Burlington area, whereas in the two previous months only one is the norm.

May usually witnesses the last killing frost of the season, generally coming between the 15th and 25th, but varying according to location. Injurious frosts have been recorded at many stations in June. The latest of record was at Bloomfield on June 30 and at Chelsea on July 1. The weather station at the latter is located in a frost hollow notorious for low temperatures and may not be representative of the surrounding countryside. The normal growing season ranges from a maximum of 162 days at Burlington in the Champlain lowland to a minimum of 110 days at Cavendish in the hills of Windsor County and 111 days at Bloomfield in the Northeast Kingdom.

May opens with a folklore holiday and closes with a national holiday of remembrance. Both are closely connected with nature's calendar. The first of May was celebrated as a spring rite in England. The community went into the woods early in the morning to gather May flowers and

49

then spent the rest of the day in feasting and celebrating the arrival of the new season. Since the 1880s, either May 30 or the Monday of the last weekend in May has been set aside to honor those who gave their lives for their country by placing flowers on their graves. One normally finds few flowers in bloom in Vermont on May 1, but the end of the month usually brings an assurance of a bountiful supply. Still, Vermont weather has many quirks. In 1903 a complete drought existed throughout May to wither nature's products, and on May 30, 1884, the day was made doubly memorable by a general snowstorm and freeze.

Historic Weather Events in May

1 1844 May Day snow: 18″ at Wheelock.
2 1884 Severe windstorm from south, shifted to west and tore away 1,100 feet of Burlington breakwater at loss of $35,000; damaged wharves at loss of $50,000.
   1903 May cold: 18° Northfield and Morrisville; Burlington 29°.
3 1790 Whirlwind observed at Rutland at about 3:30 P.M. by Dr. Samuel Williams; blew down small buildings and uprooted trees.
   1966 Cold morning: 24° at Burlington set May record there; 15° at West Burke, 18° at Bloomfield and Chelsea; 19° at Cavendish and Somerset.
4 1799 Lancaster, N.H., across river from Lunenburg, had snow a foot deep on level in woods (*History of Coos County*).
5 1812 Heavy snow near Keene, N.H.: "Snow commenced on 4th, lay 10 to 12 inches deep on high lands next morning" (*New Hampshire Sentinel*).
   1873 Brattleboro-Bennington stage on wheels instead of sleigh runners at Wilmington for first time that season.
6 1930 Early May heat wave: 92° at Brattleboro, 90° at Keene, N.H., 87° at Hanover, N.H.
7 1884 Hailfall continued several hours at Woodstock; remained on ground for over twenty-four hours; 38° at sunrise on 8th.
   1966 Cold atop Mt. Mansfield: 33°max./9°min.
8 1803 Famous May snowstorm in south; at Bennington, "it snowed considerably the greater part of the day...and continued uncomfortably cold"; heavier in southwest New Hampshire, people went to church in sleighs.
9 1966 Snowfall of 3.5″ set late-season record at Burlington; minimum 30°; Newport reported 5.2″; three snowy days, 8-10th, 11″ at Peru.

1983 May snowstorm with 4" on ground at Stowe; Rt. 108 through Smugglers Notch closed.

10 1880 "Terrific windstorm," possible tornado at Island Pond; blew down a steamboat house and covered bridge; four escaped injury although under bridge when it collapsed.

11 1907 Snow-covered mountains around Manchester; "old people say this is the most backward spring they ever knew." (*Monthly Weather Review*, Washington, D.C., May 1907).

1945 Heavy snowfall: 15" at Somerset and Wilmington, 13" at Cavendish, 10" at Dorset.

12 1870 Lunenburg: 6" snow, but no injury to plants.

1945 Record late-season temperatures: Bloomfield 24°, Chelsea and Cavendish 25°.

13 1833 May deluge at Burlington: 3.54" in 24 hours, 6.01" in four days.

1866 Tornado demolished toll bridge at Barnet; other wind damage at Middlebury and Montpelier.

14 1815 Lancaster, N.H.: "snowed almost all day," Adino Brackett diary.

15 1834 Greatest of all May snowstorms; 3 ft deep on high hills above Newbury; rate of one inch every 10 minutes for two hours; 12" reported at Randolph and Rutland.

16 1880 Tornado reported at Hanover, N.H., by *Monthly Weather Review*, but no details found in local newspapers.

17 1794 Famous frost killed early crops throughout interior New England; given historic prominence by article in *Mass. Hist. Society Transactions*.

18 1877 Tornado tore a 50-mile path, Saratoga County to Bennington County; Cambridge, N.Y., and Shaftsbury suffered damage; 5" deep hail at Bennington; very severe thunderstorm at Derby Line in north.

19 1780 Famous Dark Day in New England; noticed at sunrise at Pawlet; spread east during morning to Boston; candles needed to read; cast eerie, brassy hue on foliage and outdoor objects, attributed to smoke from forest fires in West.

1883 "The most disastrous forest fire that ever occurred in state"; originated in vicinity of Groton in southern Caledonia Co.

1982 Small tornado did structural damage at Island Pond; confirmed and photographed by Lyndon State College meteorologists.

20 1892 Strafford: "The snowstorm on the 20-21st was the greatest in the memory of the oldest inhabitants here so near to the close of the month" (*New England Meteorological Bulletin*, May 1892); surrounding hills covered up to 28"; North Bridgewater reported 30", Northfield in valley but 10".

51

21 1806 Unique dry blow: high southwest winds all day scorched leaves on trees as if they had been frosted.

22 1911 Greatest of May heat waves: 99° at Cavendish, 96° at Bloomfield and St. Johnsbury; Burlington's 92° equaled May record.

23 1955 Deluge at Burlington: 2.16" in 60 minutes, 2.23" in 120 minutes, total of 2.26".

24 1964 Early heat wave: 93° Bellows Falls, 91° McIndoe Falls.

25 1832 Snow reported 8" deep near Burlington.

26 1967 Most recent heavy May snowstorm: Northfield 9", Burlington 2.6", Montpelier 6", Rochester 2"; Mt. Mansfield 33°/27°.

27 1817 Snow fell at Williamstown, Mass., according to records of Professor Chester Dewey.

28 1775 Keene, N.H.: "Toward Night comes up a Terrible Hurricain Thunder shower. Trees are whirled down in Great Plenty" (Abner Sanger diary. Quoted by Merton Goodrich in *History of Keene*, 582).

29 1826 Snowed on height of land near Huntington (James Johns diary, Vermont Historical Society, Montpelier, Vt.)

30 1884 Randolph: "on 28th, 29th, and 30th, there were hard freezes, and on 30th there was a snowstorm that will long be remembered, as the flowers that decorate the soldiers' graves were covered with snowflakes." Lunenburg reported 2" snow and 35°. (Monthly weather report to Signal Service).

31 1843 Newbury: 3" of snow on hills.

    1961 Two-night freeze: Burlington 25° was one degree above May record; frost did $500,000 damage; West Burke 19°, Chelsea 22°, Bloomfield 23°.

June

Each spring Vermonters have to endure successive disappointments when brief periods of balmy weather are routed by a shift of wind to the northerly quadrant, making "backward spring" conditions prevail. By the time June rolls around, however, the fluctuating battle between the forces of the north and the forces of the south has usually been decided in favor of the latter, and one can go about warm season activities without trepidation.

The summer solstice comes on or about June 21 and marks the beginning of astronomical summer. The sun mounts to its highest elevation, its direct overhead rays shining down at the Tropic of Cancer just south of Key West, Florida, and Brownsville, Texas. Along Vermont's northern border the sun at noon will radiate its warmth at an angle of about 68°30', or 76 percent of the distance from the southern horizon to the zenith.

The longest day of the year occurs at this time (June 20, 21, or 22nd) with about 15 hours and 20 minutes of direct sunshine. With the sun at its most northerly latitude, twilight extends the hours of adequate visibility by almost an hour and 15 minutes at each end of the day. As a result of complexities introduced by the eccentricity of the earth's orbit and the tilt of its axis, the earliest rising of the sun comes about June 14 and the latest setting about June 27; thereafter, the sun rises later, and sets earlier, and daylight, accordingly, diminishes in duration.

Meteorological summer begins about two weeks before the solstice. The warmest consecutive ninety-three days of the year at Burlington start on June 6 and extend through September 5. This period is called meteorological summer; the mean temperature in Vermont is 63° or above.

By the opening of June the jet stream approaches its most northern location between 45° and 50°N across southern Canada, and the main storm tracks across the continent migrate northward. The most frequented route traverses Ontario and Quebec north of the St. Lawrence Valley, and storms of Alberta origination cross southern Hudson Bay instead of looping southeast to the vicinity of the northern Great Lakes. Some low-pressure disturbances form over the northern Great Plains, head northeast, and pass over Lake Superior and Lake Huron. Their trailing fronts affect Vermont with brief periods of precipitation and a change of air mass. The Atlantic coast route becomes inactive except when an early tropical storm from the Caribbean Sea or Gulf of Mexico retains its energy after passing over the land areas of the Gulf States. In June 1972 Hurricane Agnes drove north with moisture-laden tropical air and gave the Northeast a devastating deluge.

There is less anticyclonic activity in June. The main path from the Pacific Ocean crosses the continent close to the Canadian border. Over the Great Lakes the high centers are joined by a track from the Canadian Northwest, and the joint path leads east over central New York and southern Vermont. Large high-pressure areas occasionally stall over New England at this time of year and produce a series of fine days favoring outdoor activities.

Temperatures continue their upward surge, gaining an average of about 10 degrees during the month. The June mean at Newport is 62.7°, at Burlington 64.9°, and at both Bennington and Vernon 65.2°. The western section along Lake Champlain is the warmest in the state and will continue so during the entire summer. Thermometers in Vermont have dropped as low as 22° and risen as high as 101° in June.

The summer precipitation regime is in full operation in June. Thunderstorms dominate the scene. Some are of the convective type, that is, formed by rising air currents over the mountains. They develop over the central mountain spine of the state and drift eastward over the Connecticut Valley. Other thunderstorms form over the Adirondack

53

"Brookside in Late Spring" might well be the title of this Bentley photograph taken near the bridge east of the Bentley house in the Nashville section of Jericho. The young lady is Florence Durlacher and the date is circa 1925. From the W. S. Bentley collection, Jericho Historical Society.

Mountains of New York and travel across the northern counties of Vermont to New Hampshire. Many thunderstorms, especially the most violent, are associated with the passage of a cold front that lifts the warm, moist air necessary for the formation of a thunderhead. They may form in lines from north to south and subject the entire state to heavy downpours of rain, the so-called gully-washers, accompanied by loud thunder and vivid lightning.

The northern section in June is slightly wetter than either the western or southeastern. Newport has an average of 4.16 inches, while Vernon's normal catch is 3.65 inches. Some of the mountain stations report amounts in excess of four inches, lead by Peru with 4.34 inches. Dorset in a high valley in the southwest has the top figure for the state of 4.28 inches.

Historic Weather Events in June

1   1924 Cool opening: Garfield 26°, Chelsea 27°, Bloomfield and Cavendish 28°, Burlington and Bennington 35°, Vernon 40°.

1952 Torrential showers resulted in flash floods; Winooski River at Montpelier the highest since 1936. Dorset 3.95", Cornwall 3.61".

1967 Mt. Mansfield had 34" snow on ground as result of heavy spring storms.

2 1905 East Charlotte: lightning knocked several people unconscious; caused damage at East Alburg and East Montpelier.

1895 Tornado at Morrisville; path 50 rods wide; tree damage, but missed houses; accompanied by heavy hail.

3 1919 Heat wave: Burlington above 90° from 3rd to 5th; maximum 94° on 3rd.

1947 Power dam failed on East Creek at Pittsford; lower portion of Rutland flooded to depths of 9 to 15 feet; railroad bridge collapsed; Rochester 4.43", Rutland 3.08".

4 1919 Heat wave: 101° at St. Johnsbury, 100° at Cavendish; from 2nd to 6th, St. Johnsbury: 90°, 100°, 101°, 101°, 98°, on successive days.

1960 Triangular area bounded by Ludlow, Ascutney, and Chester received 3" in three hours; small streams rose to heights not witnessed since 1927; main street of Ludlow flooded.

5 1855 Destructive fires widespread; fanned by strong southwest winds, swept from farm to farm in Woodbury, burning eight dwelling houses, consuming crops, fences, forests; damage also in Waitsfield, Windsor, Worcester, Kirby, Fayston, and Berlin.

1859 Famous June frost over state; 36° at Brandon with white frost; St. Johnsbury 38°, Burlington 40°, all at 7:00 A.M., not actual minimums.

6 1816 Cold front swept southeast over state bringing historic 1816 freeze and snow in so-called "Year Without a Summer."

1925 Early-season heat wave: 99° at Cavendish, 95° at Cornwall, 94° at Burlington; generally above 85° every day from 3rd to 8th.

7 1816 Snow covered ground all day at Montpelier; on high elevations near Danville snow and sleet drifted to 20" depth; ice at Chester.

1984 Band of heavy thunderstorms passed over north-central section: St. Albans 4.45 inches and South Newbury 3.10 inches late on 6th. Severe local flooding in Washington County early on 7th; mudslide at Calais closed Rt. 14.

8 1903 Spring drought: no measurable rain at Burlington after 0.23" on April 16 until 0.15" on June 8. Yellow days occurred from 3-7th; sun appeared as red ball and foliage took on brassy hue; caused by forest fires in the Adirondacks.

9 1833 Mountains around Randolph covered with snow; flakes fell in valley.

1933 Heat wave: 96° at Cornwall, 95° at Enosburg Falls, Rutland, St. Johnsbury; Burlington 80° or more from 5-11th; 93° on 9th.

10 1816 Fourth consecutive morning with frost; severe at Middlebury with freeze, 32° at 7 A.M.

1913 Freeze: 27° at Chelsea and Somerset.

11 1842 Snowstorm over most of state: Irasburg in north 10", Bennington in south 4"; greatest amounts on high elevations, little in valleys.

1973 Tornado cut 35-mile path through Chittenden, Franklin, and Lamoille counties; damage near Milton; thunderstorm caused $300,000 other damage.

12 1847 Snow drifts still to be seen from heavy spring storms at Wheelock.

13 1833 Record downpour of 3.54" in 24 hours at Burlington, greatest in Zadock Thompson's records.

1961 Wind and hail hit Rutland Fairgrounds lifting a cattle barn 50' from its foundation.

14 1898 Heavy rain at Vernon, 3.27" in 10 hours.

15 1972 Severe local thunderstorms; Burlington 2" in 2 hours; flooding with erosion of fields.

16 1806 Total eclipse of sun; path of central totality from California to northeast Massachusetts; included Vermont south of Dorset-Chester-Springfield line; almost total at Rutland and Windsor; sky conditions perfect.

17 1967 Thunderstorms caused gulley-washers, damaging roads and utility lines; Mt. Mansfield 1.78", Rochester 0.97", Montpelier 0.66".

18 1950 Flash flood at Bennington after 4.16" on 18-20th.

1957 Tornado from Franklin to West Berkshire along Canadian border did $58,000 damage.

19 1907 Small area around Bloomfield deluged with 5.51" in 24 hours, and total of 6.82"; no other station in state had over 2".

1946 Hard freeze: Somerset 25°, Bloomfield 27°.

20 1918 Late-season frost widespread: 27° at Chelsea, 28° at Bloomfield, Northfield, Somerset, St. Johnsbury.

21 1953 Small tornado hit Champlain Country Club near Swanton, unroofed barn, knocked five-car garage from foundations.

22 1906 Heavy hail fell over area 1 x 10 miles near Chelsea: "at its height the hailstones in places were in piles from 1 to 2 feet deep."

23 1782 Tornado swarm over southern Vermont: Pawlet and Manchester hit in southwest; path from Weathersfield to Claremont, N.H., reported; hailstones measured 6" in diameter and weighed one pound at Royalton.

1940 Light to heavy frosts: 30° at Bennington, Chelsea, Somerset on 22-23rd; 31° at Bloomfield and Cavendish.

24 1816 Three-day heat wave during "Year Without a Summer": Mid-

dlebury 94° on nonregistering thermometer; Williamstown 90°, Northampton, Mass., 102°.

25 1927 Burlington's highest June wind in recent times: 32 mi/h from the south.

1983 Hailstorm with stones one inch in diameter interrupted hot air balloon event at Quechee; some balloons forced down; several people injured when struck by hailstones; about ¼" of snow fell near Sharon and Hartford, enough to make snowballs.

26 1907 Heavy rain, 1.52" in 35 minutes at Burlington; one life lost in flood.

27 1950 Heavy hail fell at Bennington having metallic nuclei; possibly carried aloft from stacks of coke plant at Troy, N.Y., to the southwest; stones 1-2" in diameter.

28 1933 Heat wave: 80° or more from 24th to 31st at St. Johnsbury; rose to 95° on 26th and to 96° on 28th.

29 1919 Late-season cold record after early-month heat wave; 37° at Burlington was date record; Northfield and Enosburg Falls hit 32°.

Haying begins in late June and, given proper growing conditions, may stretch through the summer months as farmers make two and sometimes three cuttings. Here hay is being cut and "conditioned" to speed drying on the Harold Cross farm on the Lower Elmore Mountain Road. Photo courtesy Morristown Photo Archive.

57

1946 Peak of eight-day heat wave: 96° at Burlington; 88° or above from June 24 to July 1; six days at 90° or more.

30 1973 Heaviest general rains over state since 1927; 9.80" at South Londonderry from 25th to 30th, with 6.44" on 30th; 6.50" at Searsburg; high floods on small streams.

July

"When Sirius rises with the sun, mark the dog days well begun," goes the time-honored adage for July. The brightest star of the sky, nicknamed the Dog Star, rises in conjunction with the sun for a period of six weeks from early July to the middle of August. The Romans reputedly believed that the radiation of Sirius, when added to that of the sun, caused the excessive heat of midsummer, a time when dogs were likely to go mad and people became enervated. We now know that Sirius radiates only an infinitesimal amount of heat to the earth, but the phrase "dog days" survives as synonymous with the uncomfortable temperature-humidity index conditions prevailing in high summer.

July is the warmest month of the year in Vermont, showing an increase of four to five degrees over June at all stations. Mean temperatures reach their highest point about July 20, a full month after the solstice. Thermometer readings over the different sections of the state are more uniform than in any other month: in the northeast Newport's mean is 67.1° and in the southwest Bennington has 69.1°. Several localities average slightly over 70°, such as Cornwall in the flat Champlain Valley 70.3°. The highest July temperature reading ever registered in Vermont was 105° at Vernon in 1911 and the lowest was 30° at Somerset in 1926.

From June to July a continued northward shift of most features of the general wind circulation normally takes place, and a corresponding northward displacement in the latitude of storm movement results. No main storm tracks prevail within the United States at this time, though occasional cyclonic developments may take place over the northern Plains. These move directly into Canada and pass over Quebec and Labrador well north of Vermont. The main continental storm path crosses central Hudson Bay near latitude 60°N. Trailing fronts from these disturbances may affect northern New England by bringing fresh Canadian air southward to temper a summer heat wave. Tropical storm activity increases in the vicinity of the West Indies, but few disturbances move north along the Atlantic seaboard this early in the season.

A center of anticyclonic activity prevails over the cool waters of the Great Lakes and Hudson Bay; ridges of high pressure from this activity often break away and drift eastward over Quebec and northern New

England, attended by cooling northwest winds and sparkling skies, providing some of our best summer days. In more southerly latitudes, the Bermuda-Azores high-pressure area moves north with its axis extending west over the coastal plain as far as the Appalachian Mountains. It is to blame for causing some of the extended heat waves that plague the Northeast each summer. The wide-ranging circulation around the western periphery of the Bermuda High stimulates a vast flow of southwest winds that carry heated air from the interior of the continent to the Green Mountain region.

The jet stream flows across the continent along a northerly latitude trending east-northeast from Oregon and Washington and passing north of the Great Lakes to make an exit over the Atlantic provinces of Canada.

The distribution of precipitation over Vermont in July is quite uniform, as in June. The western section receives about the same amount as the southeast, with the northern section leading both by a slight margin. Burlington averages 3.43 inches and Newport 3.95 inches. Thunderstorms are responsible for the bulk of the precipitation; Burlington reaches the season's peak with an expectancy of eight thunderstorms during the month. The mountain stations continue the wettest with about four inches of rainfall.

St. Swithin is the patron saint of weathermen, in honor of the hoary tradition that, if it rains on July 15, rain will fall for forty days thereafter. This is supposed to have occurred when the remains of the worthy Bishop of Winchester, who died in 862 A.D., were to be transferred from the churchyard to a more honored place within the church, but a long series of rainy days delayed completion of the removal. Many European countries have similar saint's days with the same tradition, but on different summer dates.

The Vermont Chapter of the St. Swithin Society does not have much positive evidence to support its reason for existence. A study of the past fifty years of records for Burlington reveals that the greatest number of consecutive rainy days following a wet July 15 is only five. In 1961, it rained on every day from the 15th through the 20th, but on three of the days only traces occurred. In 1955, four days with measurable rain followed, but all other cases were three or less. In fact, it has rained on July 15 on twenty-three of the fifty years since 1931. So it is fun to talk about St. Swithin once a year, but he really does not exert much remote control on Vermont weather.

Historic Weather Events in July

1   1863   Tornado at Bartonsville on Williams River in Windham County.
    1912   Cool opening of July: 31° at Somerset, 33° at Chelsea.

Even President Coolidge, wearing the homespun coat woven for him by his grandmother, does a little mowing on the old home farm in Plymouth while relaxing from his Presidential duties.

2 1833 Tornado traced from Salem Pond to Norton Pond in Northeast Kingdom; path at least 12 miles long; extensive forest damage.

3 1911 Famous early-July heat wave: 100° at Burlington, 98° on 4th and 5th; 101° at St. Johnsbury.

1966 Early July heat wave: 98° at Bristol, Vernon, Burlington, Bellows Falls; 97° at Morrisville, Rutland; 96° at St. Johnsbury.

4 1911 Peak of twelve-day heat wave: 105° at Vernon set all-time state maximum record; Burlington 90° or above on eight days.

5 1859 "A little frost on grass" reported at Stratford, N.H., opposite Maidstone in Connecticut Valley.

6 1897 Cloudburst of 3.68" at Chelsea in one hour during thunderstorm.

1962 Freeze at West Burke, 29°; Chelsea and North Danville 32°.

7 1777 Battle of Hubbardton, a rearguard action during retreat from Ticonderoga, fought in Rutland County on very hot, humid morning.

1984 Amtrak's *Montrealer* wrecked at Williston when small bridge was washed out by heavy overnight rains, causing sleeper train to plunge into ravine; 5 killed and 150 injured. Burlington 1.47" on 6th, Essex Junction 1.59".

8 1777 Signers of the Constitution of the new Republic of Vermont (1777-1791) were delayed in their departure from Windsor by heavy thunderstorms.

1890 Big wind over state caused structural damage and several deaths; hotel south of Plattsburg, N.Y., blown down; four drowned on lake; damage at Alburg, Hyde Park, and Montpelier.

8 1914 Flash floods at Jericho after cloudburst of estimated 8-12" fell in 1½ hours; ten bridges washed out, several buildings undermined; Burlington had only 0.26".

9 1911 Second heat wave of month at Burlington: 95°, 96°, 96° on 9-11th.

1962 Tornadoes: one path 16 miles, Chester to Weathersfield; another 8 miles, Springfield eastward across Connecticut River.

10 1961 Very heavy hail at Barre; stones 0.75" diameter; walks had to be shoveled.

11 1864 Tornado followed 12-mile path, Brattleboro to Hinsdale, N.H.; steeple of church blown down at Brattleboro; structural damage in town.

1888 Unusual July snowstorm covered mountain tops with snow following severe cyclonic storm with heavy rains and high winds; party of "twelve ladies and gentlemen" stranded atop Camels Hump overnight.

12 1898 Cool period: 34° at Northfield, 35° at Enosburg Falls on 13th; high pressure of 30.50" at Brattleboro.

13 1962 Severe thunderstorms over north; seven barns burned by lightning; 6" fell at Groton; Mt. Mansfield 4.93".

14 1897 "The great storm of the 13th and 14th blighted the good crop prospects...This storm was of exceptionally violent character for the season"; Hartland reported 7.20", Burlington 6.01"; earth slide on Fayston Mountain, 120 acres fell 2,400 feet vertically.

15 1862 St. Swithin's Day, the patron saint of weathermen; tradition that if it rains this day it will rain for forty more days does not hold in Vermont.

16 1787 Heavy hail damaged "indian corn and English grain [wheat]" at Bennington (*Vermont Gazette*).

1866 Burlington: mean of 90° was the "hottest in 40 years," *Free Press*.

17 1927 Tornado, North Troy to Elkhurst, damage $20,000.

18 1966 Long heat wave: Burlington 80° or more (daily maximum) from June 23 to July 18.

19 1850 Tropical storm moving over New York State caused southeast gales and heavy rain; Montpelier 5"; Connecticut River highest since 1801.

1953 Peak of heat wave in east: 102° at Bellows Falls, 100° at Wilder; above 90° 16th to 20th.

20 1945 "Unusually heavy rainstorm" in western Rutland County; city measured 2.91" next morning; six-day rainy spell.

21 1907 Tornado at Enosburg Falls: "thousands of sugar maples were blown over or twisted off."

1983 Line of severe thunderstorms moved across state; hailstones size of ice cubes at Brattleboro; winds in southeast at 60 mi/h; two killed.

22 1811 Great Summer Flood of 1811: struck Rutland and Windsor counties; Middletown suffered severely, all mills and dams destroyed, great field erosion and mud slides; rainfall estimate 12-15".

23 1955 After 91° temperature, thunderstorm dropped golfball-size hail at Woodstock; thunderstorms general over state.

24 1830 Heavy rain caused high flood; 3.85" at Burlington on 24th, 7" fell from 24th to 26th; Winooski River went on rampage.

25 1859 "Definitely one of the memorable floods for the Ottauquechee. Water probably higher than at any time since 1811." Woodstock newspaper clipping.

26 1859 Tornado at Groton in Northeast Kingdom, "bounced over hilly terrain," several yards wide; caused structural damage.

27 1969 Cloudburst of 6" in Ascutney area; roads washed out; communities isolated.

28 1913 Deluge at St. Johnsbury, 4.99" in 24 hours; 1.98" Chelsea.

29 1886 Heavy hailstorm at Georgia and Milton in Franklin County, wind "assumed force of a cyclone." (Note in National Archives, no author).

1969 Heavy rain in southeast: Newfane 4.18", Mays Mills* 3.98"

30 1883 Great summer aurora.

1960 Tropical storm Brenda moved on Hartford, Conn.-Concord, N.H. line with heavy rains and flooding to west; Woodstock 3", Wardsboro 2.52", Newfane 2.73", Mays Mills 3.01", Bennington 2.31".

31 1962 Cool wave ended month: West Burke 35°, Cavendish 36°, Woodstock 39°, Burlington 42°.

August

High summer, meteorologically speaking, does not come until the month of August. Though land temperatures have commenced to cool, the surface waters of the vast oceans in the Northern Hemisphere now reach their highest thermal content, and this difference exerts an important influence in steering the general circulation of the atmosphere around the globe. In August the axis of the westerly wind belt normally lies close to 50°N, and the principal storm track across North America is located at its most northerly position, near 55°N, or in the latitude of southern

*Mays Mills is located on the North River in Halifax.

Hudson Bay. Cyclonic storm centers moving across Canada may trail their cold fronts into northern New England, while southern sections of New England are not affected.

August brings "those lazy, hazy, crazy days of summer" of popular song fame. The Bermuda High continues to be a major factor in controlling the weather. On occasions its western arm may extend inland over the Carolinas and Virginia, and then heat and humidity are borne from the Gulf of Mexico on the wings of southwest and west winds to the Green Mountains. An enervating stretch of several days may ensue. One late August heat wave of southern origin began on August 28, 1973, with temperatures rising to 89° or more at Vernon in the southeast for nine consecutive days. In recent years the highest temperature during August reached 102° at Bellows Falls in 1955, and the lowest dropped to 25° at South Londonderry in 1965.

Temperatures exhibit a general decline from the July highs, but the lowering is small, about 2.3 degrees statewide. The solar input is only 85 percent of what it was in June. Bellows Falls in the southeast has the highest mean temperature with 67.8°, but this is about equal to Burlington's 67.1° in the northwest. The waters of Lake Champlain retain their summer heat throughout the month, accounting for the high figure at Burlington. In contrast, Newport in the northeast averages a mean of 64.7°, or 2.4 degrees below Burlington.

Precipitation shows a decline of 10 percent from the July averages. The wettest places are on the high elevations of the western slopes of the Green Mountains and across the northern plateau. Several stations reach above the 4-inch level. Across the north, Burlington with 3.87 inches and Newport with 4.32 inches are well above the low elevation locations in the south. The entire Connecticut Valley lies in a rain shadow, running from 3.54 inches at St. Johnsbury in the north to 3.25 inches at Union Village Dam near White River Junction in the south-central part, the lowest in August for any Vermont station.

A familiar sound on an August evening comes from the katydids who stridulantly rub their wings together and maintain a cacophonic dirge all night. The appearance of the katydid is supposed to be a warning that summer is waning and frost is in the offing. These grasshopper-like insects emerge from their eggs in late summer, lay their own batch, and pass away with the first frost of autumn, all in about a six-week cycle in northern climes such as Vermont. Though light frosts may be experienced from time to time during August, it is usually the last half of September that brings a killing frost. But there have been exceptions. The famous summer of 1816 witnessed the end of the growing season in some northern localities on August 21 when a killing frost cut down the corn, vines, and tender vegetables, and in more recent times a black frost occurred in the Northeast Kingdom on August 18, 1918, to end the shortest growing season known, only 59 days. On that date the

63

Summer time is gardening time in Vermont. The growing season varies in length from 140 days in the Champlain Valley to less than 100 days in some of the mountain townships. Late spring and early fall frosts lead gardeners to choose hardy vegetable types that will mature in the relatively short growing season. Here Lyle Griffiths of Morristown is at work in his vegetable garden. Photo courtesy Morristown Photo Archive.

temperature dropped to 32° at Chelsea, and to 33° at Bloomfield, Hyde Park, and Woodstock.

Tropical storms have become a threat in low latitudes in July, and during August even full-fledged hurricanes may move north and affect Vermont. Ex-hurricane Belle crossed the state on August 10, 1976, bringing deluges of tropical moisture to southern and central sections along its track. Many remember the winds and rains attending Hurricane Carol on August 31, 1954, as it roared northward through New Hampshire. Way back in Vermont history on August 19, 1788, a tight-knit hurricane packing tremendous power caused a tree blowdown that probably exceeded the forest devastation of the Great September Hurricane of 1938.

Historic Weather Events of August

1   1842 Cool, 50° maximum at Randolph: "kept fire in office all day," William Nutting weather diary. (Snow fell at Quebec City on 1st and 2nd.)

1964 Earliest freeze and frost in modern records: 32° at Chelsea and Somerset; 30° at South Londonderry on private thermometer.

2 1975 "Hot Saturday" during heat wave; 100° at Cornwall, 99° at Burlington, South Hero on the 1st; 99° at Vernon, Chelsea on the 3rd; all-time New England high of 107° in Massachusetts.

3 1970 Waterspout on Lake Champlain, moved inland at St. Albans along five-mile path.

1923 Plymouth: Calvin Coolidge took the oath of office at 2:27 A.M. under favorable weather conditions; clear at sunset, 5/10ths of sky covered by clouds at sunrise at Northfield, the nearest weather station; 63° overnight low; no appreciable wind.

4 1833 Frost at Windsor and wood fires necessary; snowflakes seen at Woodstock.

5 1881 Heat wave: 102° at Bradford and Hanover, N.H.; 100° at White River Junction.

1955 Record high August wind at Burlington, 54 mi/h from the north; 95° max./70° min.

6 1856 Excessive rains covered southwest corner; 11.80" to 18" in thirty-six hours; damaging flood at Bennington and vicinity.

7 1918 Hottest day of intense two-day heat wave: 102° at Vernon, 100° at Cornwall, 99° at Cavendish.

1955 Flash flood at St. Albans after 1.48"; Bloomfield 3.52" on 6-7th; Enosburg Falls 2.23", Newport 1.95".

8 1965 Man killed by lightning while in construction shack at Mt. Holly.

1972 Damaging thunderstorm at Burlington; hail 0.75" diameter; structural and tree damage from possible tornado.

9 1829 "Tremendous Whirlwind" struck Peacham and Barnet; destroyed barns and orchards; unroofed houses; path 10-35 rods wide.

1976 Heavy rains from passage of ex-hurricane Belle on 9-10th; Readsboro 4.11", Mt. Mansfield 4", Dorset 2.83"; local flooding, soil erosion, and landslides near Manchester.

10 1892 Lightning destroyed chair-making factory at Rutland; loss of $75,000.

1949 Burlington hit 97° at climax of five-day heat wave.

11 1944 Burlington at all-time high of 101°, peak of long heat wave; 89° or more for nine consecutive days, 9-17th.

12 1887 Frost noticed in low places at Lunenburg on 12th and 13th.

13 1955 Ex-hurricane Connie dumped heavy rains over south: Wardsboro 5.10" in 24 hours, total 7.86"; Whitingham 8.29", Weston 8", Readsboro 8.08", South Londonderry 7.97"; worst floods since 1927 and 1936 resulted.

14 1840 Strong northeast gale and heavy rain swept western counties; 1.38" rain fell at Burlington.

15 1960 Lightning at Windsor killed a Morgan horse, a former national grand champion.

16 1777 Battle of Bennington: heavy rain delayed opening action, enabling militia to make forced march from Manchester and arrive in time to join in afternoon battle, insuring victory.

17 1955 Deluge of 3.59" at Burlington in 24 hours.

18 1918 Frost in low places; killing at Bloomfield with 33°; Chelsea lowest at 32°; Hyde Park and Woodstock 33°.

    1969 Turbulent weather: thunderstorm outbreak across north; 2" hailstones reported; waterspout on Lake Champlain.

19 1788 Unique minihurricane with small central core crossed southeast corner; destructive wind lasted only 20 to 30 minutes; enormous forest destruction; many cattle killed.

20 1950 Flash flood on Walloomsac River in Bennington: 1.05" on the 18th, 2.39" on the 19th, 0.72" on the 20th.

21 1925 Record August cold: 26° at Somerset equaled August minimum record for state; Garfield 30°, Cavendish and Chelsea 31°.

22 1960 Heavy rains: 1" in 20 minutes near Halifax; 2" in less than one hour at Springfield.

23 1915 Tropical storm on wayward track over lower Great Lakes and down St. Lawrence Valley produced heavy rains: Vernon 3", Somerset 2.61", Bloomfield 1.48", Chelsea 1.44".

24 1933 Tropical storm moving on Washington, D.C.-Binghamton, N.Y.-Burlington line caused heavy rainfall: Cavendish 2.86", Woodstock 2.70", St. Johnsbury 2.58", West Burke 2.62", Burlington 2.16".

25 1884 Heavy frost; fields of buckwheat at Bennington completely ruined; corn, beans, and vines killed at Lunenburg.

    1982 Deep low-pressure system moving over Northeast dropped barometer at Burlington to 29.28" for record August low pressure.

26 1913 Freeze: 30° at Somerset and Chelsea.

27 1856 Three inches of snow atop Mt. Washington in New Hampshire.

    1948 Nine-day heat wave at Bellows Falls, 90° or more from 24th to 30th, peak 98° on 26th and 27th; four other stations hit 98°.

28 1971 Tropical storm Doria crossed state with heavy rains: Mt. Mansfield 4.50" plus 1.16" on 29th; Cornwall 3.69", Cavendish 3.14" on 28th; washouts on several principal highways.

    1973 Lightning hit ski lodge in Warren killing two persons.

29 1893 Second tropical storm in a week moved northeast over New York State causing gales and heavy rain over Vermont; Burlington reported 1.49" with a southerly wind flow; wind damage reported several places.

30 1957 Early freeze: 30° at West Burke, 31° at Bloomfield, 32° at Chelsea.

1965 Mt. Washington reported 1.6" snow on 29th, 0.9" on 30th, a modern August record; Mt. Mansfield had maximum temperature of 38° and minimum of 28° on 30th.
31 1932 Total eclipse of sun, cutting across Northeast Kingdom from Derby Line to Bloomfield; partially obscured by middle and high clouds.
1965 Record three-day cold spell at end of August; 25° at South Londonderry, 26° at West Burke, 27° at Northfield and Chelsea.

September

The arrival of the autumnal equinox on or about September 23 marks a significant event on nature's calendar. In the astronomical realm, it indicates that the direct overhead rays of the sun have crossed the equator in their annual southward migration, resulting in a steadily declining solar input for the Northern Hemisphere. The North Pole has seen the last of direct sunlight, and the pall of darkness descends the ladder of latitudes each day. For the atmosphere over North America, it means the doldrum-like circulation of summer is at an end, and storm movement across the continent will soon quicken. A new climatological season is at hand, and all living things must make adjustments to the changing weather conditions.

Having a lower angular slant, the sun's rays must penetrate a greater slice of the earth's atmosphere and now impart only 65 percent of the radiant energy to the surface in Vermont compared with the high input at the summer solstice three months before. At the end of the month, the noontime sun stands at 46° or about halfway from the horizon to the zenith, having declined about 11 degrees during the course of the month. The period of direct sunlight shortens noticeably, by about 40 minutes at the end of each day.

The jet stream makes its most northerly entrance into the continent in September, then pursues an easterly course close to the 49°N parallel, crossing Lake Superior and southern Quebec and leaving the continent over the Atlantic provinces of Canada. The major storm tracks, too, continue at their northerly positions. Cyclonic activity is mainly confined to Canada where Alberta-type storms (*see* Glossary) move east between 55°N and 60°N across Hudson Bay and northern Quebec. Cold fronts trailing from the storm centers introduce cool polar airstreams into the Green Mountain region, permitting temperatures to drop to the freezing mark and below with increasing frequency as the month progresses.

September brings the peak month of hurricane activity. Large tropical storms originate off the coast of West Africa and make the long journey across the broad Atlantic Ocean to the West Indies. They then either con-

67

tinue west into the Caribbean Sea and Gulf of Mexico or recurve to the northwest and then north along the Atlantic seaboard of the United States or over the offshore waters. The most recent examples of this type that affected Vermont were the Long Island-New England Hurricane in September 1938, Donna in 1960, and David and Frederic in 1979.

The principal September anticyclone tracks resemble those of August. A well-defined primary path extends in a zone across the northern tier of states near parallel 45°N. The northern plains lead all areas in the frequency of high-pressure systems, which blocks storm generation in the middle of the country. An early portent of autumnal conditions appears in the renewal of high-pressure generation in the intermountain region of the West, whence a secondary track leads east to a center of growing importance in West Virginia and the central Appalachian area.

Mean temperatures decline by about 7 to 8 degrees in the course of September, and a chill in the morning and evening atmospheres becomes quite noticeable. Newport has a mean of 56.7° and Vernon of 60.5°. Thermometers in recent Septembers have ranged from a high of 100° at Bellows Falls and Vernon in 1953 to a low of 15° at Dorset in 1947. Burlington has experienced a smaller range: 95° to 25°.

Precipitation in September shows a distinct decline from the high figures of the summer months. The influence of thunderstorms in swelling rainfall totals diminishes; on the average Burlington experiences only three thunderstorms this month compared to eight in July. The southeast becomes the wettest section, mainly the result of tropical storms passing along the coast and extending their rain shield to the near portion of Vermont. Stations in the mountains average slightly over 4 inches: Mays Mills 4.08 inches, Searsburg Station 4.37 inches, and Whitingham 4.23 inches. The driest localities lie in or near the Northeast Kingdom: Montpelier 2.90 inches, Gilman 2.99 inches, and St. Johnsbury 3.06 inches. September can be a droughty month when a renewal of cyclonic storms coming from the west is delayed and tropical storms fail to move northward near enough to the seacoast to influence Vermont skies.

Historic Weather Events in September

1 1843 Mt. Moosilauke in New Hampshire reported covered with snow as seen by David Johnson from Newbury.

    1953 Four-day August heat wave extended into September; 89° or above at Burlington from 1st to 5th, with 90° on 2nd, 3rd, 4th.

2 1967 Cool morning: West Burke 30°, Bristol and Cavendish 32°, Burlington 38°; six-day cool wave at West Burke with less than 40° every morning from 1st to 6th.

3 1904 Thunderstorms did extensive damage; two large barns in Ben-

Wilson S. Bentley was interested in photographing all kinds of weather phenomena, not just snowflakes. Among his several cloud studies was this one taken up the hill behind his Jericho home. From the W.S. Bentley collection, Jericho Historical Society.

son hit by lightning and burned; residence in East Bethel met same fate; two barns in East Fairfield hit, one burned; minor damage in Vergennes, Stowe, Bristol.

4 1828 Heavy rains of 9.70″ at Newfane on 2nd-4th caused a "most destructive freshet at Windsor." (General Martin Field records in National Archives)

    1973 Hottest day of extreme heat prevailing Aug. 28 to Sept. 5; 95° at Vernon, 89° or above from 1st to 5th; 92° at Morrisville on 5th.

5 1963 Frost and freeze: Chelsea 28°, West Burke 30° on 5th and 6th; 32° at Cavendish.

6 1881 Famous "Yellow Day" throughout Northeast; smoke from Michigan forest fires filtered sun's rays; all verdure took on a brassy hue; artificial lights needed in late morning; affected all parts of state.

    1888 East Berkshire: "The frost on the 6th was exceptionally destructive to corn, buckwheat, oats and potatoes which owing to the backward season were unusually green, and in many instances in this vicinity were ruined"; 27° on 6th, 30° on 7th. (*New England Meteorological Bulletin*, Sept. 1888)

69

6 1979 Remnant of Hurricane David moved northeast across central section with winds up to 50 mi/h; widespread power outages; light to moderate crop damage; heavy rain but little flooding; Pownal 3.77", Dorset 4.15", Searsburg Station 3.66"; lightning destroyed barn near Bethel; possible small tornado near Rutland.

7 1953 Stagnating low pressure trough caused very heavy rains in Northeast Kingdom: West Danville 4.33", Bloomfield 3.10", Lemington 2.58", West Burke 2.83", St. Johnsbury 2.64", all within 24 hours on 6-7th.

8 1945 Late heat wave: Burlington 94° on 7th and 8th, 90° or above 6th to 9th.

9 1821 Great New Hampshire Whirlwind: apparently originated in Windsor, crossed river into Cornish, N.H., pursued path across Lake Sunapee, over Mt. Kearsarge; six killed; most massive tornado in New England until Worcester Tornado in 1953; other smaller tornadoes that afternoon in Vermont at Pittsford and Berlin and at Haverhill, N.H., opposite Newbury.

10 1913 Freeze and frost: 27° at Chelsea and Somerset.

    1955 Highest wind ever in September at Burlington: 48 mi/h south.

11 1931 Late heat wave set a September record, 95° at Burlington; 95° at St. Albans, 93° at St. Johnsbury, 92° at Northfield.

    1968 Heavy rains at Fairlee and Vershire washed out beaver dams, flooding and destroying roads; Cavendish 3.53", Union Village 3.40".

12 1960 Extensive Hurricane Donna crossed southeastern New England; pressure below 29" over state except in northwest; heavy rains: Woodstock 4.61", Wardsboro 6.03", Somerset 6.33", Readsboro 5.27", Newfane 5.76".

13 1948 Late heat wave: Burlington and St. Johnsbury 92°.

    1964 Burlington 32°, its earliest freezing temperature.

14 1882 Tornado at Strafford; path 4 miles long, 0.25-mile wide; buildings unroofed, trees uprooted.

    1979 Cold front associated with remnant of Hurricane Frederic caused severe thunderstorms; 80 percent power outage in central section; wind damage in Addison and Orange counties; bridge over Connecticut River at South Newbury a principal victim; many trees downed in Chittenden and Franklin counties; 72 mi/h gust at St. Albans.

15 1859 Hard frost at Lunenburg, 23°; very short growing season after June 12 frost.

16 1932 Tropical storm crossing Cape Cod produced heavy rains in Connecticut Valley, almost none in Champlain Valley: Bloomfield 2.70", Brattleboro 2.24", but only 0.06" at Burlington.

    1939 Burlington reached 92°, its latest 90° reading ever; East Barnet

96°, St. Johnsbury 95° during record late heat wave from 15th to 17th.

17 1948 Cool wave: 24° at Dorset and Somerset, 25° at Lemington, 27° at Woodstock, 28° at Bennington.

18 1908 Smoky skies from forest fires during week ending on the 18th; very dry month: Northfield only 0.35″ with none from 4th to 28th.

19 1827 Very heavy rains at Newfane, 8″ fell on 19th to 21st, according to General Martin Field's weather records.

20 1845 Great Adirondack Tornado pursued path of about 275 miles from south shore of Lake Ontario, over lake as waterspout, through Adirondacks, across Lake Champlain as spout, into South Burlington where it caused structural damage.

21 1938 Great New England Hurricane cut diagonally across state from southeast near Brattleboro to northwest near Burlington; heavy rains on 18-19th plus hurricane deluges on 20-21st; totals, Somerset 8.20″, Brattleboro 7.76″, Woodstock 5.74″; high winds, Burlington 47 mi/h south; damaging floods; vast forest blowdown.

22 1918 Snow fell in the White Mountains and on Mt. Mansfield; 36° at Northfield and St. Johnsbury, 44° at Burlington.

23 1815 Center of Great September Gale passed directly north over central New Hampshire; heavy rains and floods mentioned in Vermont press, but no wind damage since Vermont lay in less severe western section of storm.

   1885 Earliest general heavy snowstorm of record: 8″ to 10″ reported by weather observers at Randolph and Strafford; more on mountains.

24 1963 Very cold: 15° at Chelsea, 19° at West Burke, Woodstock; Burlington's 25° equaled September record.

25 1950 Cool morning: West Burke, Cavendish 22°; Dorset 24°, Woodstock 25°.

26 1950 Blue Sun and Moon observed from 25th to 29th; smoke from forest fires in western Canada filtered rays, depleting green and yellow, permitting passage of blue; temperatures reduced, growing season ended.

27 1816 Black frosts killed crops over entire state on 27th and 28th; unripened corn destroyed; Hanover, N.H., reported readings of 23°, 20°, 20°, 25° from 26th to 29th.

   1947 Very cold: Dorset 15°, lowest ever in New England in September.

28 1836 Snow reported covering the mountains of southern Vermont by observer at Williamstown, Mass.; first of three very early snows.

1962 Unusual secondary center of coastal storm formed over western Vermont; high winds did extensive damage; heaviest losses in Bennington and Rutland counties; two radio stations knocked out; little rain fell.

29 1844 Six inches of snow fell at Wheelock in northern Caledonia County.

1914 Burlington's 25° equaled September record; 20° at Northfield and Enosburg Falls.

30 1835 Great September Snowstorm of 1835 covered north; 6-12" in Franklin, Orleans, and Essex counties. (Zadock Thompson reported the details while residing at North Hatley, P.Q.)

October

The tempered weather of October makes it the favorite outdoor month of the year across most of America, and nowhere is it held in greater esteem than by the residents of Vermont. The Green Mountains attract more one-day visitors in bright October than in any other month. The annual autumnal migration of foliage viewers is the direct result of the changes set in motion by the further decline of direct solar rays to a mid-month position over the Amazon Valley of Brazil. The diminished heat input causes the chlorophyll of the deciduous trees to cease functioning, resulting in green leaves transforming to a brilliant variety of reddish and yellow tints bedecking hill and valley. The mixture of flaming hard-woods with deep green pines creates striking contrasts across northern New England found nowhere else in the United States.

October is the time to expect the first spell of Indian summer, which some people refer to as a "second summer." By tradition, Indian summer must follow the first frosts of autumn. The latter have been staged by an anticyclone from Canada bringing down cold air on the wings of northwest winds. When the center of the high pressure area has passed to the east, the winds swing around to the southwest and blow gently, introducing a period of mild days with hazy sunshine prevailing—Indian summer.

The sky takes on a hazy appearance around the horizon and smog condition may envelop urban areas. Temperatures rise into the high 50s and low 60s and one feels the urge to be out-of-doors and resume some of summer's activities. Indian summer is a sort of floating season on the calendar and may occur anytime from mid-October to early December. There may be only one such period in an autumn, or two or three visitations, the sequence varying from year to year.

Mean temperatures decline about 10 degrees in the course of October, although the gradient between north and south remains relatively small: Newport has an October mean of 46.1° and Vernon 49.4°. The absolute extremes over recent years have been from a maximum of 91° at Vernon in 1963 to a low of 5° at Morrisville in 1972. The October range at Burlington can be from 85° to 17°.

Precipitation shows a substantial decline of about eight percent from the September totals, the greatest amount falling in the western section and the least in the southeast. Of a representative group of valley stations, Rutland receives the least amount of rain with 2.70 inches, while Vernon has 3.49 inches. The mountain stations report more than four inches: Searsburg Station 4.35 inches and Peru 4.26 inches.

Historic Weather Events in October

1   1959 Tropical storm Gracie, moving along southern border, produced heavy rains: Waitsfield 2.64", Cornwall 2.50", Bethel 2.39", Peru 2.20".

2   1908 Drought eased by first substantial rains since August 17; Burlington reported 0.64"; Northfield had only 0.35" all September.

    1947 Late-season heat wave: Rutland 87°, Vernon 89°.

3   1780 Pomfret: "a light snow about three inches deep," Jonathan Carpenter diary.

4   1869 October 1869 Flood, greatest of the nineteenth century; Ottauquechee River was 2' higher than September 1828; heavy rains on 3rd turned into deluges as tropical moisture arrived with Saxby's Storm offshore; 6.57" Castleton, 6.35" Woodstock.

5   1965 Wintry period: snow fell on 4th, Enosburg Falls 3", Newport 1.1", traces at many stations; cold day on 5th, Mt. Mansfield max. 19°; Burlington 37°/31°.

6   1836 Second snowfall in a week; Bennington had snowflakes mixed with rain, but Manchester Mountain all white on 7th. Camel's Hump and northern peaks also covered.

    1932 Heavy rains: Burlington 4.21" in twenty-four hours; St. Albans 4.31"; Cornwall 4.05".

7   1962 Coastal storm augmented by moisture from offshore Hurricane Daisy produced heavy rains over Vermont: Danville 5.52", Woodstock 4.23", St. Johnsbury 3.90", Newfane 3.82".

8   1965 Winds funneling down narrow valley of Huntington River at an estimated 90 mi/h in gusts caused considerable damage to trees and homes; storm windows and doors blown off; four large trees in Huntington Center uprooted.

9   1888 Wet period: fourteen consecutive days with measurable precipitation at Burlington ended.

73

10 1804 Snow Hurricane of 1804; tropical storm moving northeast over southern New England caused gales, thunder, heavy precipitation; press estimated snow depths 36" to 48" fell on higher elevations; *Farmers Cabinet* at Walpole, N.H., thought 15" to 18" accumulated in valley, but much melted; enormous damage to orchard trees.

10 1925 Famous October Weekend Snow of 1925; up to 24" reported in Stowe Valley with entire state having some; Burlington 4.3", Northfield 5"; highways had to be plowed; many football games cancelled.

11 1869 Second heavy rainstorm within a week brought rivers to near-record crests again, though not much left to be damaged; Randolph 2.35".

12 1836 Third early-season snowstorm of 1836; ground covered at Bennington with 6" and temperature 33°; Huntington and Hanover, N.H., had 5".

13 1934 Early snowstorm: 8.2" at Bloomfield, 5.7" at West Burke.

14 1846 Energetic hurricane moving north over Pennsylvania caused strong southerly winds over state; these funneled through valleys at high speed; widespread minor damage reported, but no rain.

15 1820 Connecticut River at Windsor rose 10 feet in 4 hours, and high waters closed all bridges over the Onion (now Winooski) River in north.

1954 Hurricane Hazel moving north over Pennsylvania caused only structural and tree damage; Burlington maximum wind 70 mi/h from the southeast.

16 1955 Second period of excessive rains over south only two months after deluges by hurricanes Connie and Diane; stalled front and secondary center south of Long Island produced 8.65" at Mays Mills on 15-16th and 11.77" on 14-20th.

17 1780 Famous Indian raid on Royalton took place with 8" of snow on ground.

1947 Heat wave: Burlington's 85° equaled October record; Bellows Falls hit 88°, Chelsea 87°, St. Johnsbury 86°.

18 1783 Second tropical storm on New England coast caused heavy snowfall over state; 12" reported at Woodstock; probably deeper on mountains.

19 1902 Thunderstorm with high winds and lightning caused damage at Underhill, Richford, Barton, Hyde Park — large losses by individuals.

20 1823 Early snowstorm: 8" at Newfane; stage on runners at Barnard.

1972 Record early cold wave: 15° set Burlington's all-time October minimum; 5° Morrisville, 9° Woodstock on 21st.

21 1926 Early snow: 7" at Woodstock, 6" at Hanover, N.H.

22 1783 Melting of previous snowfall plus rains "raised streams to such a degree as produced the greatest flood ever known since the settlement of this country." *Vermont Journal* (Windsor).

1791 Snowfall of 10" reported at Rutland by Dr. Samuel Williams.

1969 Snowstorm: Montpelier Airport 8" with 12" on high hills. Rochester 11.1", Burlington 5.1" in 24 hours (October record).

23 1785 Great New England rainstorm caused widespread floods; Newbury had highest water experienced prior to 1850, oxbow intervales along the river submerged.

24 1959 Heavy rainstorm: Mays Mills 5.20", Vernon 3.84", Woodstock 3.75"; washouts occurred.

1969 Early cold: 9° at Cavendish and Rochester, 10° at West Burke.

25 1963 80° at Burlington, latest reading in the eighties in season records.

26 1912 Somerset received 6.51" rain, 23rd to 26th, resulting in floods.

1925 Second snowstorm of month occurred when heavy rain ended in snow to cover ground; Somerset 2.20" total precipitation.

27 1823 Very heavy early-season snowstorm of 22" reported at Newfane by General Martin Field.

28 1965 Early-season snowstorm: 7" at Enosburg Falls, 3.2" at Montpelier Airport.

29 1952 Snowstorm: 11.5" at Bloomfield, 10.5" at Lemington, 7.2" at St. Johnsbury, 6" at Northfield; only trace at Burlington.

30 1866 Heavy rains caused freshet at Wilmington; water said to be highest in fifty years on Upper Deerfield River.

31 1925 Coldest October ended on a cold note; 2° at Garfield set state October record; Bloomfield 6°, Enosburg Falls 8°.

1965 High winds swept state; wall of new high school at Duxbury collapsed.

November

The pace of atmospheric activity across the continent quickens in November. The frequency and severity of cyclonic storms increases, and their main path moves closer to the Green Mountain region. Precipitation becomes more frequent, and the amounts are heavier. Cloudiness reaches a maximum for the year, sunshine is at a minimum, and general gloomy conditions pervade atmospheric Vermont. Now it is often cold enough to snow, and the landscape can be transformed overnight as by magic from the stark gray of late autumnal forests into a winter wonderland of white.

In November the main jet stream, after crossing the middle of the Pacific Ocean near 40°N, enters the continent over Vancouver Island, and then races east in undulating fashion with a sweeping curve into the Midwest as far south as the Ohio Valley. Trending northeast, it carries over central New England where it is a principal agent in creating the storminess and heavy precipitation that characterizes the eleventh month in Vermont.

Several new features of the winter storm tracks make their appearance this month. Low-pressure systems generate over the eastern slopes of the Rocky Mountains, mainly in Alberta and Colorado, and converge over the Great Lakes to create a zone of maximum storminess there.

The St. Lawrence Valley serves as the pathway to the northeast for low-pressure systems, which cause Vermont's weather to experience quick changes as the warm and cold fronts sweep through. An additional storm track from the Gulf of Mexico sends cyclonic disturbances northeast with increasing frequency along either flank of the Appalachian Mountains, both paths ultimately affecting Green Mountain weather.

The winter path of polar anticyclones from northwest Canada becomes well established during November. Moving southeast over the western prairies and northern plains, the polar anticyclones pass west and south of the Great Lakes. Later they often stall over the central Appalachians and give rise to a West Virginia high situation that blocks the normal movement of storms across the country and brings on a spell of settled conditions over the Northeast.

The downward trend in temperature accelerates from October to November, the decrease ranging from 12.6 degrees at Newport in the northeast to 11.5 degrees at Dorset in the southwest. The absolute difference of monthly means between north and south increases to about four degrees: Newport 34.3° and Vernon 38.5°. The extremes over recent years run from 81° at Bellows Falls in 1950 to −19° at White River Junction in 1938. The range at Burlington has been from 75° to −3°.

Precipitation increases from October to November by moderate amounts in the northeast and west, but in the southeast section November leads all other months for wetness, having an average of 4.23 inches, or 1.02 inches more than October. The larger amount can be attributed to the greater frequency of storms, especially along the Atlantic seaboard where northeast coastal disturbances develop and sometimes attain unusual severity. Vermont's greatest rainstorm, resulting in the tragic Flood of 1927, was a November product of the subtropical Atlantic Ocean. On the average the drier northwest has a November catch of 2.80 inches at Burlington, in contrast to the wetter south where Searsburg Station has 5.18 inches, a ratio of almost one to two.

Thanksgiving is a holiday anticipated and enjoyed by all, but often it brings far from ideal conditions for going "over the river and through the woods to grandmother's house." The fourth Thursday may fall on any day from the 22nd to the 28th. A study of weather conditions at Burlington reveals that temperatures during this period have ranged from 69° in 1953 to −3° in 1938. On 53 percent of Thanksgiving Days, no measurable precipitation has fallen, and no heavy falls of rain over 1 inch

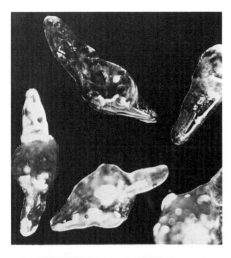

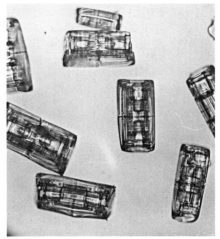

Frozen water may take many shapes as these W. S. Bentley photographs clearly show. The ordered beauty of the snowflake design contrasts with the irregularity of sleet (left) and the barrel-like formation of ice crystals (right).

have been recorded in the past century. Snow fell on 73 percent of the precipitation days. The biggest Thanksgiving Day snowstorm occurred in 1971 when as much as 20 inches fell atop Mt. Mansfield and 16.3 inches at Newport. The snowfall covered all the state, with Vernon in the southeast reporting an even 12 inches.

Historic Weather Events in November

1 1716 Dark Day (caused by forest fires to the west): "In America, the 21st October (Old Style) or November 1 (New Style) was so dark people used candles," Noah Webster, from Cotton Mather.

77

1951 Moderate snowstorm: 6" St. Johnsbury and St. Albans; 5" Enosburg Falls and Dorset.

2 1819 A famous Dark Day in Vermont and Canada; candles needed to see in daytime in Champlain Valley; caused by smoke from forest fires.

1950 Heat wave set date records of 71° on 1st and 75° on 2nd at Burlington; 79° at Rutland, 78° at Cavendish, Dorset, Woodstock.

3 1950 Heat wave continued in east: Bellows Falls 81°, Wilder 78°, but Burlington's maximum only 45° west of cold front.

4 1927 Vermont Flood at height. Total rainfall: 9.65" at Somerset, 9.14" at Molly's Falls, 8.66" at Northfield, 8.60" at Chittenden, 8.47" at Rutland; greatest damage in Winooski and White River valleys; most bridges carried away; much railroad track destroyed; highways washed out; fields covered with gravel; eighty-four dead; damage estimated at $28 million.

5 1955 Early snowstorm: 10" Windham County; 29 major telephone lines downed by wet snow.

6 1948 Late heat wave: Cornwall 77°; Burlington's 75° set late-season and November records; Newport in northeast hit 71°.

1951 Cold opening week: minimum below 32° every day from 1st to 7th at Burlington, lowest 13° on 6th; 6° at Dorset and Enosburg Falls.

7 1908 Lake Champlain was 2⅝" lower than at any time since 1827 after long summer-fall drought.

1968 Heavy snowfall of 10-20" on high elevations permitted earliest opening of ski trails; extensive damage occurred to utility lines in valleys.

8 1938 Climax of eight-day warm spell; Rutland 70° or higher from 5th to 8th, 76° peak on 7th; Burlington and St. Johnsbury, 74° on 7th.

9 1819 Another dark day (caused by forest fires) described in article by Prof. Frederick Hall of Middlebury; called the "dark year" at Lunenburg.

10 1948 74° set late-season heat record at Burlington; 75° Enosburg Falls, 74° Cornwall and St. Albans.

11 1955 Snowstorm of 3-5" on high ground caused postponement of many Veterans Day games due to slippery highways.

12 1899 Minimum of −5° at Enosburg Falls was earliest below-zero reading of record: St. Johnsbury 3°, Woodstock 4°, Hartland 5°, Derby 6°.

1968 Third major snowstorm dropped another 24" on high ground; Newport 10.8", Peru 10", Enosburg Falls 7".

13 1933 Dust Bowl cloud from the Great Plains reached the state for the first time, reducing visibility.

14 1972 Big snowstorm with heaviest amounts in the south; Peru 13.5", Readsboro 12.6", Cavendish 10.5"; worst damage to utility lines

since 1938 hurricane; Burlington only 1.7" with 35° max./25° min.

15 1933 Mid-November cold wave; Burlington 9° on 15th, 6° on 16th; Woodstock −6°, Rutland −5° on 16th.

16 1906 Early snowstorm at Burlington dropped 13" with 32°/25°; good sleighing at Groveton, N.H., after 18.5" fell.

17 1869 Strong southwest and south winds western half of state; railroad cars blown from tracks at Chatham, N.Y.; estimated a "strong gale," or equivalent of 60 mi/h, at Plattsburgh, N.Y., on Lake Champlain.

1977 Heavy rains, high winds, and hail during storm; lightning struck radio station atop Mt. Mansfield; 0.32" on 17th, 1.15" on 18th.

18 1964 Gust of 62 mi/h at Burlington knocked down many limbs and wires.

1965 Heavy snowstorm across north enabled ski resorts to make early start; Enosburg Falls 14.0", Newport 12.8", Burlington 8.8".

19 1921 Latest 70° or more readings: Enosburg Falls 75°, Burlington and Cornwall 70°; highest east of the mountains was 66°.

20 1836 Red snow fell over northwest part of state.

1869 Second big wind in three days swept the hill towns along the western slope of the Green Mountains; extensive structural damage with valley locations suffering, too; deep low-pressure center in Quebec caused strong southerly winds.

21 1900 Violent wind for 20 minutes at Hartland in afternoon "did much damage to property in this vicinity, blowing down trees, barns, and other buildings"; damage also reported at Enosburg Falls and Stratford, N.H.

22 1931 Six-day heat wave hit peak of 67° at Burlington; after 66° on 21st, 65° on 23rd and 24th.

23 1943 Great November snowstorm in Northeast Kingdom: Bloomfield 31.5", Newport 29.5"; Northfield 22.4".

24 1938 Winter's start: heavy snowstorm on 23-24th followed by moderate fall on 27th gave large monthly totals in southeast; Newfane 23.5" with over 20" at Brattleboro, Bellows Falls, Mays Mills, Searsburg, Somerset.

25 1938 Earliest below-zero reading for season at Burlington: min. −1°, max. 14°.

1950 Great Appalachian Storm: gales swept state for many hours; peak sustained speed of 72 mi/h from south at Burlington; extensive structural damage.

1971 Big Thanksgiving Day Storm: 20" Mt. Mansfield, Waitsfield, and Montpelier; wires down and roads blocked.

26 1784 Shirkshire windstorm in Bennington Valley did extensive damage; details reported in *Vermont Gazette* during its first year of publication.

1938 After early-month heat wave, late-month cold wave descended: −19° at White River Junction, −17° at Chelsea and Cavendish, −16° at Woodstock.

27 1883 Spectacular sunrises and sunsets as result of dust veil aloft from Krakatoa volcano in East Indies: "western sky lighted up as if by a gigantic fire in the heart of the Adirondacks." (*Monthly Weather Review*, Washington, D.C., Nov. 1883)

1898 Heavy snowstorm occurred over Vermont while the famous Portland Storm was ravaging the New England coast: 20" at Chelsea, Hartland, and Woodstock.

28 1921 Severe icestorm in south, heavy snow in north; Bennington 2.23", Somerset 3.04", Cavendish 2.24"; Bloomfield had 40.2" snow in November.

1984 Temperature soared to record 68° to climax a week of late-season warmth.

29 1875 "A gale-y day," *Free Press* described "one of the severest days known for years; Newport dropped from morning 32° to evening −8°, and to −18° next morning with strong northwest wind blowing.

30 1875 Coldest November day of record at Burlington: maximum −1°, and minimum −10°.

1963 Record low November barometer of 28.62" at Burlington; gales and heavy rains swept southern part; Vernon 1.76", Newfane 1.55".

December

"Dark December," the apt phrase from Shakespeare's *Cymbeline*, well describes the concluding month of the year in Vermont. The sun on December 21 stands at 21°33′ above the southern horizon in central Vermont compared to an altitude of 69°27′ on June 21, and the hours of direct sunlight decline to about nine hours a day compared to about fifteen and a half hours at the summer solstice. The solar beam at the end of December radiates about 21 percent of the energy it put forth over the Vermont countryside six months ago. Cloudiness, too, adds to the general gloomy atmosphere. Normally at Burlington, twenty-one of the thirty-one days are rated as cloudy, that is, as having more than 60 percent cloud cover. An all-time record for any month was established in December 1954 with twenty-six cloudy days.

The main Pacific jet stream follows about the same path as in November, except over the eastern states it continues directly eastward from the Ohio Valley and passes well south of Vermont. An additional jet stream pursues a path over the southern United States, then sweeps northeast to

The snowstorm of December 2, 1936, left Burlington's Church Street adrift. City trucks and crews came to the rescue the next morning, but note that in this depression year when labor was plentiful, the trucks were loaded by hand. A McAllister photograph courtesy Bailey/Howe Library, University of Vermont.

the Carolina capes, and thence over the offshore waters to join the northerly jet stream east of Cape Cod. The principal storm tracks from the west migrate south of their November positions, with some storm centers crossing the Green Mountain area. With increasing frequency, barometric depressions move along the Atlantic seaboard as coastal northeasters whose precipitation can encompass all of New England with heavy rain or snow.

Anticyclonic movement takes on a winter pattern. The main track from the Canadian Northwest continues to introduce arctic outbreaks of frigid air that ultimately reach Vermont in modified form, and a new secondary center of anticyclonic activity is found over the Province of Quebec. In conjunction with a center of minimum mean pressure off Cape Cod, an active air flow often crosses New England from northwest to southeast.

The decline in mean temperature is more evident from November to December than between any other months. The amount ranges from about 13 degrees in the southeast to about 15 degrees in the northeast. Newport returns a December mean of 19.4°, compared to Vernon's 25.4° and Bennington's 26.1°. Over recent years the absolute extremes in December have been from −50°, the all-time Vermont minimum recorded at Bloomfield in 1933, to 72° at Enosburg Falls in 1941. The range of temperature in December is the greatest for any month of the year.

81

Precipitation drops off from the high levels of November by almost 20 percent in the northeast and west and by 13 percent in the southeast. Amounts in the mountains about equal the summer high figures: Mays Mills averages 5.21 inches and Searsburg Station 5.11 inches in December. Valley weather stations record much less precipitation, as exemplified by Champlain Valley locations where Burlington's average is only 2.43 inches, indicative of the predominance of dry westerly winds from the interior of the continent.

One expects winter to be in full swing by Christmastime, and Vermonters are seldom disappointed in this regard. The chance of having a white Christmas with one inch or more on the ground is 77 percent at Burlington in the Champlain Valley, and the figure rises to 93 percent at higher elevations in the northeast, and probably to 100 percent atop the peaks. On Christmas Day in the past, Burlington has experienced temperatures as high as 62° in 1964 and as low as −25° in 1980. Few will forget that recent bitter holiday, when the maximum of −5° occurred just after midnight, and at 3:00 P.M. the thermometer stood at only −14°. Nor will memory fade of the great Day-after-Christmas Snowstorm in 1969 when a record 29.8 inches fell at Burlington and 45 inches at Waitsfield. So great was the problem of clearing the highways that Governor Deane C. Davis declared a state of emergency in order to obtain federal assistance in opening the roads for holiday travelers.

Historic Weather Events in December

1 1842 Big snowstorm at Randolph: 24″ on Nov. 30-Dec. 1, William Nutting record.
   1934 Burlington set a December maximum reading to that time with 64°.
2 1942 Deep cyclonic storm dropped barometer at Northfield to 28.45″, to 28.49″ at Burlington; Northfield max. wind 30 mi/h from the south; heavy rain in south: Newfane 1.98″, Vernon 1.88″.
3 1936 Snowstorm at Burlington of 11″ on 2-3rd; temperature range from 29° to 13° on 2nd, 31° to 25° on 3rd.
4 1940 Very cold first six days set early-season records: Enosburg Falls −34°, Chelsea −31°, Bloomfield −29°, St. Johnsbury −27°, Burlington −17°.
5 1941 Late heat wave sent thermometer to 72° at Enosburg Falls, highest ever in Vermont in December; Burlington 67°, Rutland 66°, Bennington 65°.
6 1951 Very warm opening of December; maximum above 58° at Burlington from 5th to 8th; peak of 64° at Enosburg Falls on 6th; Dorset 65°, Rutland 64° on 8th.

1981 Heavy snows started one of longest skiing seasons; Mt. Mansfield 20″ on 6-7th, Peru 18″, West Burke 13″, Burlington 10.2″.

7 1950 Burlington had southeast wind at 47 mi/h.

1959 Southwest gales caused widespread minor damage; branches felled powerlines; shingles ripped from roofs and sidings.

8 1902 Early-season cold: Enosburg Falls had maximum of −5° and minimum of −20°; Newport −17°; Northfield −15°.

9 1902 Early cold wave intensified: Newport −26°, Wells and Jacksonville −24°.

1963 Thunderstorms reported over the south with an inch of precipitation.

10 1966 Warm period set three consecutive date records with 61° maximum on 9th, 10th, and 11th at Burlington.

11 1958 Unusual cold wave of long duration; West Burke below zero every day from 8th to 27th; coldest: −29° on 11th, −30° on 12th.

12 1952 Snowstorm on 12-13th, 13.4″ at Burlington; 24″ at St. Albans, 24″ at Enosburg Falls, 15.8″ at Newport, 15″ at Lemington.

13 1915 Pre-winter heavy snowstorm: Burlington 12.8″ with 15″ depth on ground; Somerset 2.82″ precipitation.

14 1901 Warm midmonth: 63° was highest so late in season at Burlington; reached 62° on 15th.

1920 Big wind from southeast at Burlington at 47 mi/h, but no precipitation.

15 1839 Famous December 1839 storm struck all New England; numerous shipwrecks on coast, heavy snow interior; 12″ fell at Randolph in "violent snowstorm"; 2 ft reported at Chester; no mail for many days.

16 1835 Vermont's bitterest daylight period; northwest gales swept entire state; temperature remained near −15° all day in all sections; minimum of −22° reported at Newbury that night.

17 1951 West Burke reached nadir of −29° during prolonged cold wave from 13th to 21st with the minimum below zero each day; −30° at Barnet.

1973 Snowstorm dropped 12.8″ at Burlington; St. Albans 10″, but only 4″ at Vernon.

18 1967 December thunderstorm at East Richford struck transformer damaging wiring and causing power outages at Newport and Highgate.

19 1961 Wet snow and icestorm on 17th to 20th; from 6″ to 14″ snow fell; power lines down, four-day outage at Mays Mills.

20 1942 Burlington's −23° set early-season record; −32° at Newport and Bloomfield, −29° at St. Johnsbury.

21 1905 Shirkshire winds in Bennington Valley. At Manchester, "it was one of the worst known here, several buildings were unroofed,

83

windows broken and barns doors blown off," W. W. Canfield, weather observer.

22 1850 Heavy snowstorm at Randolph on 22-23rd, 18" fell.

1955 Severe cold wave; Burlington −22° on 21st, −20° on 22nd, −18° on 23rd, all date records; West Burke down to −33°.

23 1929 Heavy snowstorm at Burlington, 14.5".

1970 Blizzard conditions with heavy snow; from 8" to 16" fell over state; Montpelier 12.5"; West Burke −25°: many pre-Christmas activities cancelled.

24 1850 Big Christmas Eve snowstorm during record snowy December; Burlington measured 21" on 24-25th.

1966 Heavy snow fell across the state, a boon to holiday skiers; 16" St. Albans Bay, 14.9" Burlington, 10.5" Vernon.

25 1872 Coldest Christmas: −45° at Lunenburg; said to be −54° near Stratford, N.H.; "coldest spell since January 1844"; −41° at Woodstock, −40° at Randolph.

1964 Warmest Christmas: 62° at Burlington set date and late-season record; Dorset 67°, Bennington 66°, but cold on east side of mountains where Vernon had 39° maximum.

1980 Bitterest Christmas Day: Burlington fell from a midnight reading of −5° to a next-midnight reading of −25°; Mt. Mansfield dropped to −38°; snow fell across state, with a maximum of 4.5" at Enosburg Falls.

26 1969 "A devastating northeaster" commenced in memorable post-Christmas storm; up to 45" fell at Waitsfield; Burlington had its biggest single snowstorm with 29.8" from 26th to 28th; governor declared emergency.

27 1866 Famous Berkshire Storm extended into Vermont, "worst in 20 years," dropped 25" at Wilmington; gales caused drifting and blocked highways for days.

28 1884 Amazing temperature rise of 49 degrees at White River Junction from −25° on 27th to 24° on 28th.

29 1917 Famous year-end cold wave from 29th to 31st; Northfield had minimums of −32°, −41°, and −41°; St. Johnsbury, −22°, −43°, and −41°; Burlington, −23°, −25°, and −19°.

30 1933 Lowest official temperature ever reported in Vermont: −50° at Bloomfield; Burlington had barometer of 31" and record low reading of −29° on 29th and 30th, both records for these dates.

31 1948 Wet end of year: excessive rains on 30-31st caused floods, especially on western side of mountains; Searsburg Mt. 7.52", Somerset 7.32", Dorset 6.70", Bennington 6.09".

1963 Cold New Year's Eve: −29° at Woodstock, −28° at West Burke, −19° at Burlington.

1965 Warm New Year's Eve: 60° at Dorset, 59° at Bennington, 56° at Burlington and Montpelier; cool east of mountains.

1816: "A YEAR WITHOUT A SUMMER"

"Eighteen hundred and froze to death" was a familiar expression a century ago whenever New Englanders gathered to talk about the weather. A chance May snow flurry or a frosty morning in June stimulated fears that another cold growing season might be at hand, such as their parents and grandparents experienced in their youths.

Many accounts of varying authenticity appeared in almanacs and local histories concerning "a year without a summer," and they usually pinpointed the reference to the year of economic stress following the conclusion of the War of 1812. One can read all sorts of tall tales about the strange antics of the weather elements in 1816, even spurious reports of snow on the Fourth of July and heart-rending stories about ancestors starving on their farms in the midst of barren fields. No subject in the weather history of New England arouses so much interest today as does the Summer of 1816, so a study in depth as to what actually happened in the Green Mountains at this time deserves special consideration.

Was there actually "a year without a summer," when "no month passed without a frost, nor one without snow," and when "no crops at all were produced"? The search for an answer presents an involved but fascinating task for the historian, leading not only into local meteorological and climatological sources, but into the economic and social realms to assess the immediate and lasting effects of the unusual season on the Vermont community.

The amount of source material for the study of Vermont weather during the year 1816 certainly is not prolific, but appears adequate for our task. Although the scientific study of meteorology was in its infancy in the early nineteenth century, certain individuals kept records of their observations of local weather conditions on their own initiative; their notations can be analyzed to give a picture of the actual weather conditions that prevailed across New England during this period.

A complete record of daily data is available for Williamstown, located in extreme northwest Massachusetts adjoining the Vermont community of Pownal, where Professor Charles Dewey began a valuable series of fifteen years extent in 1811. Readings of temperature, wind direction, and sky conditions were made three times a day, along with daily precipitation and useful comments about unusual events.

A monthly abstract of the daily weather record kept at Middlebury College by Professor Frederick Hall has been preserved in printed form, though the original manuscript has not been located. Its data are helpful in establishing the monthly range of temperature and departures from normal for locations in western Vermont. Though a thermometer was located at Hanover, New Hampshire, in 1816, presumably at Dartmouth College, no continuous record of the daily readings has been preserved. Temperature readings at other locations in the Connecticut Valley can be gleaned from newspaper accounts of the coldness of the season.

For a regional view of the daily weather map developments – the passage of cyclones and anticyclones and the arrival and departure of cold and warm fronts that affected Vermont weather – the records maintained at Brunswick, Maine (Bowdoin College), Cambridge, Massachusetts (Harvard College), and New Haven, Connecticut (Yale College), as well as those at Williamstown and Middlebury, combine to provide sufficient data to reconstruct the daily weather map developments. The gathering of the reports from this "college network" was coordinated and published by Professor Frederick Hall at Middlebury in the *Literary and Philosophical Repository* that he edited at this time.

Other very valuable sources are the diaries of individuals who were usually engaged at least part-time in agricultural work and were especially sensitive to the trend of the weather elements. Benjamin Harwood at Bennington provided a day-by-day account of the local weather events and their effect on his pursuits during the summer of 1816, as did Adino Brackett, a farmer and schoolmaster of Lancaster, New Hampshire, located across the river from Lunenburg, Vermont. Excerpts from other diaries appearing in many town histories enrich the story of the Summer of 1816 with local details.

The arrival of each untoward weather event and its influence on the progress of the crop season can be followed in the columns of the weekly newspapers of the state. Several complete files for the year 1816 have been preserved either in the original newspaper edition or on microfilm. Ebenezer Eaton in the *North Star*, published at Danville in Caledonia County, supplied valuable accounts of the weather and crop season at his northern frontier outpost. Other journals of special value in this study are the Bennington *Green Mountain Farmer*, Burlington *Vermont Centinel*,

Some account was given . . . of the unparalleled severity of the weather. It continued, without any essential amelioration, from the 6th to the 10th instant — freezing as hard five nights in succession as it usually does in December. On the night of the 6th, water froze an inch thick — and on the night of the 7th and morning of the 8th, a kind of sleet or exceeding cold snow fell, attended with high wind, which measured in places where it was drifted, 18 or 20 inches in depth. Saturday morning the weather was more severe than it generally is during the storms of winter.

North Star
Danville, Vt.
June 15, 1816

Montpelier *Vermont Watchman*, Windsor *Vermont Journal*, and *Rutland Herald*. In the Connecticut Valley, the pages of the Keene *New Hampshire Sentinel*, Walpole *Farmer's Museum*, and Hanover *Dartmouth Gazette* supplied pertinent details.

Climate Setting

The early years of the nineteenth century lie within the period of history known to paleoclimatologists as the Little Ice Age. — spanning the years from about 1650 to about 1890 when a downward swing in temperature trends produced averages about two to three degrees Fahrenheit colder than in the present century. Temperature records for North America exist only for the latter part of this cold period, and the data points to the decade from 1811 to 1820 as the coldest ten-year period, and to the years 1815 to 1818 as having the greatest departures from normal.

The best continuous record in the United States during the early nineteenth century was maintained at New Haven, Connecticut, on or close to the Yale College campus. The only years in the first half of the century to have an annual average below 48° were 1812, 1815, 1816, 1817, 1818, 1821, 1835, 1836, 1837, and 1843. As we said, the decade from 1811 to 1820 was the coldest with an average of 47.82°, and the four years from 1815 to 1818 averaged 46.78°. It was even colder from 1835 to 1837 with an annual average of 46.07°, but the remainder of the decade turned relatively warm. No other combination of two or more consecutive years averaged below 48° except those from 1815 to 1818 and from 1835 to 1837.

Professor Hall at Middlebury described 1812 as a year when "crops were destroyed by the coldness and wetness of the season." His maximum temperature in June 1812 reached only 83° and in July, 87.5°, or about 10 degrees below the current expected maximums for these months.

His rain gauge caught a total of 7.7 inches in June and 5 inches in July, or about 175 percent above normal. Though no frosts occurred in the early summer, the thermometer dipped to 37° in August. Because this reading was taken at 7:00 A.M., one could speculate that frost probably formed around sunrise at susceptible locations.

Adino Brackett at Lancaster, New Hampshire, also took note of the discouraging season of 1812:

> July 29—From the 15th of June to this time we have had a continued series of wet and cold weather. Grass is good, as is wheat, but corn is very backward.
>
> August 22—Today it rains and has rained except for one day for a fortnight past. We have had only five fair days since we began to hay (on July 10). There never has been such a continued rain before to my knowledge.
>
> August 29-30—A considerable frost.
>
> September 20—Began to snow at 4:00 A.M. and continued till after sunset. Ground white next morning and hill tops had snow all day.
>
> September 27—A hard frost sufficient to kill all kinds of vegetables.
>
> September 28—Another hard frost. Corn very poor all over N. England.

Fair crops were grown in 1813 and 1814, though the temperature lagged below normal during most of the growing season. The spring of 1815 proved backward with snows and frost in mid-May and early frosts in late summer shortening the growing season. Adino Brackett recorded the general conditions experienced by many farmers in the Green Mountain State during the summer of 1815:

> Through the whole of July the weather continued warm and growing. It seemed as though old times had returned. But the first day of August a great change took place.
>
> August 9—Last night we had a frost which on the hay looked quite white. It collected on the scythes in considerable quantity as we were mowing in the morning.
>
> September 4—Almost every vegetable was killed by the frost last night. It was heavier on the meadows than on the hills, there being no fog. So severe a frost so early in the year I have never known.

Winter and Spring, 1816

The winter of 1815-16 was not severe. In fact, it rated the mildest during the cold decade of 1811-20. Noah Webster at Amherst, Massachusetts, not far from the Vermont border, related in his diary: "The winter was open, a snow in January, which was sufficient for sledding, was swept away in a few days. The ground was uncovered most of the winter." In the northern Connecticut Valley at Lancaster, New Hampshire, Adino Brackett described the season: "the last was a moderate winter. Snow not more than two feet deep at a time. The first of March was very warm and almost all the snow went off." The lowest temperatures reported in Vermont over the winter of 1815-16 were above the usual minimums expected: Williamstown (Mass.) $-13°$ and Middlebury $-19°$.

The trend toward consistently below-normal temperatures did not appear until mid-spring. April, running only slightly below normal, ended with a heat wave sending the mercury to early summer levels: $82°$ at Middlebury and $80°$ at Williamstown, Massachusetts. About this time reports from different localities in New England mentioned spots on the sun and its reddish appearance, "giving a gloomy aspect, not unlike what is noticeable in the time of a solar eclipse, when one half of the sun's disk is obscured." This was later attributed to the presence of some type of dust high in the atmosphere. The same condition had been noticed by Benjamin Franklin in the winter of 1784 when a number of untoward meteorological events occurred.

Temperatures in May did not respond in normal fashion to the increasing solar altitude. Snow fell in eastern Massachusetts on May 10, and the roofs of houses at Albany, New York, were covered on the 14th. Professor Hall at Middlebury reported heavy frosts on successive nights from the 15th to 17th and noticed snow on the Green Mountains. With cloudy days and cool nights prevailing, the flowering season ran well behind schedule. The best phenological records for New England at this time were kept at Mansfield in southeast Massachusetts, where the Stearns family had recorded the progress of the seasons since 1798. The apple blossom date in 1816 on May 28 was the second latest in their records; only the cold spring of 1812 was later. The New Haven, Connecticut, record confirmed this.

June's Advent

Prospects of the agricultural community were dismal as the month of June opened, but spirits revived momentarily when a heat wave of midsummer proportions spread over the Northeast on June 5. Thermometers rose above $90°$ over southern New England that afternoon, and at Williamstown in northwest Massachusetts the mercury climbed to $83°$ at noon before checked in its rise by an afternoon thunderstorm. Observer Chester Dewey described the conditions as "very hot & sultry morning and forenoon."

At this time, New England lay in a tropical airstream on the eastern

89

side of a low-pressure trough that extended southward from a cyclonic storm center, probably located over the Province of Quebec. Thundershowers occurred in advance of the eastward-moving cold front imbedded in the trough of lowest pressure. One can make a plausible surmise of the weather map situations for these June days by assembling the temperature, wind direction, and barometric readings of the stations of the college network located in different parts of New England.

On the afternoon of June 5th, while showers were in progress in the tropical airstream over most of New England, the cold front of the cyclonic weather system entered northern Vermont and began a sweep to the southeast that was to make economic and social history, as well as to establish meteorological records.

When the cold front arrived at a locality, the wind shifted rather abruptly from south to northwest, thunderstorms rumbled across the sky, the heat was quickly routed, and thermometers tumbled precipitously as cold airstreams arrived on the scene. Temperature descents of as much as 49 degrees by the morning of the 6th occurred in eastern New England where the maximums had been in the 90s the previous afternoon. At Williamstown, Massachusetts, the thermometer read 45° at 7:00 A.M. and did not rise above that mark during the rest of the day, described by Professor Dewey as a "cold rainy day, wind N.W. – not much rain – wind high and very chilly."

As the low-pressure system moved eastward off the coast into the Atlantic Ocean, surges of cooler and cooler air poured into New England from a cold source region in central Canada. When temperatures at two to three thousand feet aloft dropped to the freezing point, the light rain changed to light snow over the highlands of Vermont. Valley locations in the west and south witnessed only a few wet flakes, but northern sections and higher ground generally throughout the state received a white covering of varying depth. Snow was light on the 6th, but another influx of cool air introduced by a secondary cold front arrived late on the 7th. This was attended by snow in substantial quantities over the plateau of northern Vermont.

The editor of the *Vermont Journal* at Windsor in the central Connecticut Valley took notice of the unusual weather events:

> The Weather. – The extraordinary weather which has been experienced during the present season must afford much matter of speculation to the meteorologist. The inhabitants of every part of the U.S. have remarked on the uncommon variableness or uncommon drought. In this neighborhood on Mon. last [June 3], it was clear & cold; on Tues. forenoon the same, and in the afternoon showery and sultry. Thurs., cold and squally, with a sprinkling of rain, and some snow which was quite visible on the trees of Ascutney Mt. Fri. clear, except flying clouds, and so cold, especially in the forenoon, as to require almost winter fires. Sat. [June 8]

still uncomfortably cold, squally and blustering through the day—mostly cloudy—winter fires and winter groups around them. Considerable snow has fallen in this state and N.H. We are informed that snow has fallen in Randolph and other parts of this state, several inches deep.

Probably no one living in this country ever witnessed such weather, especially of such long continuance.

Vermont Journal (Windsor), June 10, 1816.

In Vermont's northernmost press, the *North Star* of Danville, editor Ebenezer Eaton described even more severe conditions of that early June:

Melancholy Weather. Some account was given in the last issue of the unparalleled severity of the weather. It continued, without any essential amelioration, from the 6th to the 10th instant—freezing as hard five nights in succession as it usually does in December. On the night of the 6th, water froze an inch thick—and on the night of the 7th and morning of the 8th, a kind of sleet or exceeding cold snow fell, attended with high wind, which measured in places where it was drifted, 18 or 20 inches in depth. Saturday morning [the 8th] the weather was more severe than it generally is during the storms of winter. It was indeed a gloomy and tedious period. The shoots and leaves of forest trees, which were just putting forth, and corn and garden vegetables that were out of the ground, were mostly killed—But it is thought corn will recover from the shock, and that the buds and blossoms . . . are injured very little, if any. Since Monday [the 10th] the weather has been tolerably good; the citizens are recovering their spirits, and the din of industry is again heard.

The events of the week and their effect on the work schedule of farmer Benjamin Harwood of Bennington were set down in his voluminous diary covering local happenings from 1805 to 1837:

June 6. It had rained much during the night and this morning the wind blew exceedingly high from NE raining copiously, chilling and sharp gusts. About 8 A.M. began to snow—continued more or less till past 2 P.M. The heads of all the mountains on every side were crowned with snow. The most gloomy and extraordinary weather ever seen.

June 7. In ploughfields and other parts the surface of the ground was stiff with frost—the leaves of the trees were blackened—past 6 in the morning a wash-tub full of rain water was scum'd with ice, snow remained on Sandgate and Manchester Mountain past noon or as late as that. Wind extremely high night & day and the cold abated but little in the P.M. Father & Mr. Brown rode till noon hunting sheep—mended fences with greatcoat & mittens on.

June 8. The awful scene continued. Sweeping blasts from North all the forepart of the day, with light snow squalls. More clear P.M.—no snow—wind not so high—but held cold. Were principally engaged in digging stone. So cold in the morning that we

91

were absolutely compelled to send for our mittens and wear them till near noon-day.

June 9. Frosty morning, perfectly clear all day, dry chilling N wind.

June 10. Another frost, cold day, indeed obliged to thrash our hands while hoeing. Corn, which had been up a few days badly killed—difficult to see it—gloomy weather.

June 11. Frosty morning, but fine day.

June 12. Warm & smoky—signs of rain.

James Winchester related an experience when he was a fourteen-year-old boy in Vermont in 1816.

An uncle of mine had some sheep in a back pasture lot. To get to that lot he had to go through a piece of woods for nearly a mile. The weather had been very cold all June. The big storm of the 17th [7th?] began along about noon, and my uncle started after dinner to go out to the sheep pasture to fix up a shelter of some kind for the sheep. No one had any idea, cold and eccentric as the season had been up to that time, that we would have a fall of snow that would amount to anything at that time of the year. I was at my uncle's when he left home to go to the sheep lot, and as he went out of the door he said to his wife in a jocular way: "If I am not back in an hour, call the neighbors and start them after me. June is a bad month to get buried in the snow, especially when it gets to be so near the month of July."

Nothing more was thought of the matter. The snow increased in fury, and by night it had drifted so that the roads were almost impassable, but even then, and when it grew dark, none of us felt uneasy about uncle. The weather had become bitter cold. When night set in in earnest and there was no sign of my uncle's return, his wife sent me and my cousin, who was two years younger than I, to alarm the neighbors and tell them that we believed that uncle had been lost in the snow and had perished.

We had a hard time getting to the nearest neighbor's, less than a mile away, and there we gave the alarm, but could go no further. The neighbor summoned others, and in spite of the severity of the night, they searched the woods until morning, but no sign of the missing man could be found.

The search was taken up by them the next day and all of the next night without any trace of him being discovered, except that he reached the pasture and built a shelter of boughs in one corner of the lot, under which the sheep were huddled. On the forenoon of the third day the searchers found my uncle buried in the snow a mile from the pasture, in almost an opposite direction from home. He was almost frozen stiff.

He had evidently become bewildered in the blinding storm and had wandered about until he succumbed to fatigue and cold. It seems a most improbable thing that a person ever fell victim to a snow storm in the middle of June in this latitude, but I have this

sorrowful knowledge of one instance where such a thing was only too true.

The appearance of the warm spell in September, dispelled that fear for a time, but on the 16th of the month [26th?] the cold weather returned suddenly, and the calamity believers were once more made miserable by their old fear. One old man, James Gooding by name, was so hopeless over the prospect that he killed all his cattle and then hanged himself, after vainly trying to induce his wife to make away with herself also, to escape the terrible and gradual death by freezing and starvation, which he believed was to be the common doom.

The amount of snow varied, both with latitude and elevation. The greatest depths were reported on the northeast plateau where diary quotes in town histories mentioned as much as 12 inches at Craftsbury, though the five to six inches reported at Lunenburg and Lancaster, New Hampshire, were more typical of the northern section. The ground remained white all day on the 8th at the state capital at Montpelier after an overnight fall, and substantial amounts drifting to one foot were seen on the surrounding hills.

In the western valleys, only a few flakes were observed at Burlington and Middlebury. Farther south at Rutland, high elevations were white for awhile, and the mountain tops and upper slopes down to Bennington were covered with snow of undetermined depth. At Williamstown, Mass., it snowed several times during the day with both round snow pellets and regular flakes falling, but the ground was not whitened as at surrounding towns with higher elevations.

East of the mountains as much as four inches were reported at Randolph, and the town historian of Barnard spoke of five to six inches. On the floor of the Connecticut River Valley from Hanover southward, only mixed rain and snow fell with no covering. The trees on the slopes of Mt. Ascutney near Windsor were white, according to the editor of the *Vermont Journal*. No snow reached as far south as Amherst, Mass., though Noah Webster saw some on the hills in Vermont to the northwest.

Temperatures reached freezing almost everywhere on the morning of the 7th. The thermometer at Middlebury read 32° at 7:00 A.M. and was probably slightly lower at sunrise two hours before. Chester Dewey at Williams College in extreme northwestern Massachusetts had 35° at 7:00 A.M. and reported both frost on the ground and ice on water in an open bucket. Ice formed to the thickness of an inch at some northern locations.

The cold relaxed slightly on the morning of June 9th, but the 10th again brought a hard freeze. Chester Dewey's thermometer at Williamstown, Massachusetts, was "below 32°" at sunrise according to his notes, and the area experienced "a severe frost...killed beans, corn, cucumbers, etc."

Clear skies, permitting an intense overnight radiation of heat to outer space, brought an even lower temperature of 30.5° at 5:00 A.M. on the 11th, but by early afternoon the air warmed to 70° on the urging of a southerly wind. The unprecedented early June cold spell of six days duration came to an end.

Normal June weather returned after the 11th with maximum temperatures rising into the 70s and 80s for the next ten days. A heat wave of midsummer intensity prevailed from the 22nd to the 24th, culminating in a 94° reading at Williamstown on the final day. The Middlebury thermometer in the usually warmer Champlain Valley reached a maximum of 99° in June, presumably on one of these three days. In eastern New England, the instrument at Salem, Massachusetts, rose to 101° and at Waltham, a suburb of Boston, successive daily maximums attained 93.5°, 99°, and 98°. So Vermont and the rest of New England did have a sample of summer-type weather, even in 1816. Despite the heat burst late in the month, June averaged well below normal in all parts of New England: Williamstown, Massachusetts, by 3.2 degrees and Cambridge, Massachusetts, by six degrees Fahrenheit.

July 1816

The month of July was marked by an absence of warm nights so necessary for corn development. At Williamstown, Massachusetts, only ten mornings had 7:00 A.M. readings with the mercury standing at 60° or higher and only one with 70° or higher, indicative of the absence of tropical air with its overnight warmth and humidity. Wind flow came from the northwest for 50 percent of the observations, bringing successive influxes of chilly Canadian air. During a cool four-day period early in the month widespread light frost formed: at Middlebury on the 8th and

JULY hath 31 days, 1816.

First quarter, 2d day, 4 h. 37 m. morning.
Full moon, 9th day, 7 h. 43 m. morning.
Last quarter, 17th day, 7 h. 53 m. afternoon.
New moon, 24th day, 6 h. 17 m. afternoon.
First quarter, 31st day, 9 h. 34 m. morning.

M.	W.	Calendar, &c.	☉R.	☉S.	☽'sp.	r ☽s
1	2	Sultry hot weather with	4 24	7 36	♎	11 54
2	3	thunder and lightning.	4 24	7 36	♎	Morn.
3	4	Sun slow clock 4m.	4 25	7 35	♎	0 23
4	5	Independence U. S. '76	4 25	7 35	♏	0 52
5	6	Battle at Chippewa 1814	4 26	7 34	♏	1 18
6	7		4 27	7 33	♐	1 56
7		4th Sunday after Trinity	4 27	7 33	♐	2 34
8		☽ S. 11h 22m. ☽ lo. in s.	4 28	7 32	♑	3 17
9		Fairfield burnt 1779	4 28	7 32	♑	☽ rises
10		Columbus born 1448	4 29	7 31	♒	8 2
11	5	Cloudy and perhaps	4 29	7 31	♒	8 49
12	6	Day 15 hours long rain	4 30	7 30	♒	9 36
13	7	very hot	4 30	7 30	♓	10 3
14	F	5th Sund. after Trinity	4 31	7 29	♓	10 30
15	2	with	4 32	7 28	♈	11 2
16	3	Stoony Pt. F. tak. 1779	4 33	7 27	♈	11 33
17	4	thunder	4 34	7 26	♈	Morn.
18	5	lightning	4 35	7 25	♉	0 1
19	6	and perhaps	4 36	7 24	♉	0 36
20	7	rain.	4 37	7 23	♊	1 12
21	F	6th Sund. after Trinity	4 38	7 22	♊	1 52
22	2	Moon high in the north	4 39	7 21	♋	2 41
23	3	Sun enters ♌ Rain	4 40	7 20	♋	3 32
24	4	is expected	4 41	7 19	♌	☽ sets
25	5	B. at Bridgewater 1814	4 42	7 18	♌	8 21
26	6	at this time	4 43	7 17	♍	9 0
27	7	Dog days begin	4 44	7 16	♍	9 26
28	F	7th Sund. after Trinity	4 45	7 15	♍	9 57
29	2	Dog Star south 10h.	4 46	7 14	♎	10 21
30	3	Good weather	4 47	7 13	♎	10 53
31	4	for the farmers.	4 48	7 12	♏	11 23

The Vermont Register and Almanac for the Year of Our Lord 1816 *predicted "sultry hot weather with thunder and lightning" for the first days of July, 1816, and "very hot" weather toward the middle of the month. While almanacs have never been trustworthy weather guides, the predictions for 1816 couldn't have been wider of the mark.*

at Williamstown and Windsor on the 9th, but this caused little or no harm to crops. Light frosts also occurred in the upper Connecticut Valley, according to the diary of Adino Brackett of Lancaster, New Hampshire. The Middlebury thermometer dropped to a July low of 34°, and a 7:00 A.M. reading of 43° was noted at Williamstown, Massachusetts.

As the month progressed, a new threat to the growing crops appeared in the form of drought. At Middlebury no rain, "except for a few drops," was recorded by Professor Hall from June 15 to July 17. At Williamstown during the same period only 1.13 inches were measured and 0.84 inch of this fell on June 24-27. The July catch there totaled 2.14 inches, or only 59 percent of the contemporary July average. The historian of Stowe stated that snow fell on the upper slopes of Mt. Mansfield during every month save one in 1816—presumably July.

August 1816

The first nineteen days of August brought a return to near-normal midsummer temperatures, although it continued very dry. Thermometers climbed into the low 80s during the second and third weeks, capped by readings of 92° at Middlebury and 90° at Williamstown on the 18th and 19th. A circulation from the warm interior of the continent prevailed, and the hopes of the agricultural community revived. "The weather for some time past has been very favorable for vegetation, which perhaps never increased so rapidly in growth in the same space of time," declared the editor of the *Farmer's Cabinet* at Amherst in southern New Hampshire.

Unfortunately, the favorable turn in the weather was not destined to continue long. The same editor in his next weekly issue observed: "A great and sudden change in the weather...took place on Tuesday [the 20th]. About noon a violent storm of wind and rain went over. It came up very suddenly and was of short duration, but it rained and blew tremendously accompanied by heavy thunder. After which the air became cold." The arrival of the cold front took place about the same hour at Williamstown in northwestern Massachusetts where the thermometer dropped from an early afternoon 71° to an evening reading of 49°.

When the skies cleared and the winds calmed on the evening of the 21st, a heavy frost formed overnight. Observers reported its presence at Middlebury in western Vermont, at Orford in the upper Connecticut River Valley of New Hampshire, and at Williamstown in northwestern Massachusetts, so presumably the entire state had a white visitation. The temperature at the latter read 36° at 5 A.M. on the 22nd.

After a short respite, the cold-bearing winds from the northwest returned again on August 27th and brought an even heavier frost on the mornings of the 28th and 29th. Across the northern part of Vermont and

New Hampshire the corn was killed. According to Samuel Morey at Orford, New Hampshire, and the editor of the *New Hampshire Sentinel* at Keene (bordering Windham County): "The severe frost of Wednesday [the 28th] has put an end to the hopes of many corn growers. Some whole fields have since been cut up for fodder."

No alleviation of the drought appeared. The Williamstown, Massachusetts, rain gauge caught 2.13 inches in July and only 1.69 inches in August, or 61 percent of the contemporary normal. The only rainfall came from light showers attending the cold front passages. No tropical air masses were present to provide copious afternoon showers, nor did any general storm spread over the region, so constantly did the wind circulation hold to a northerly or westerly quarter.

The *Boston Centinel* in late August reported for the "information of meteorologists and weatherwisers…in Vermont there are complaints of uncommon cold, and that the mountains are covered with snow." The snow cover reference must have been to mountain tops only as no evidence of a general snowstorm in August in 1816 has been found for the Green Mountain region.

September 1816

The same general weather circulation continued into September with temperatures remaining at below-normal levels and precipitation continuing scant. On only one September day did the thermometer mount into the 80s at Williamstown (Mass.); most morning readings were in the 50s. A cool spell prevailed from the 12th to 14th with early afternoon temperatures remaining below 60°. Snow was reported on the higher peaks at this time, though none fell in the valleys.

September's rainfall catch of 1.10 inches at Williamstown amounted to only 30 percent of the contemporary normal, and brought the total figure for the late summer, July to September, to only 49 percent of normal. Tropical moisture finally reached southern New England at mid-month when a vigorous coastal storm came north and produced a six-day rain period. Heavy amounts were measured: five inches at New Haven and seven inches at Albany, New York. But the precipitation edge did not reach far into Vermont. Williamstown, Massachusetts, measured only 0.56 inch over three days. The continued drought in northern New England caused fires to break out in the tinder-dry woods. Chester Dewey at Williams College reported a smoky atmosphere from the 24th through the end of the month.

The climatic event of the unusual season came with black frosts during the last week of September. The severity of the successive freezes may be judged by the sunrise readings at Hanover, New Hampshire: on the

26th, 23°; on the 27th, 20°; on the 28th, 20°; on the 29th, 25°. The correspondent of the *Dartmouth Gazette* reported: "These frosts have destroyed all the corn, and the potatoes are much cut off by the drought and frost." Williamstown also reported freezing temperatures in September and "severe frost" from the 26th through the 30th, with a minimum temperature of 25° at 7:00 A.M. on the 28th. Middlebury registered a low of 32° at the same hour. The hard freeze ended the growing season and any future agricultural prospects over all of New England from the Canadian border to Long Island Sound.

October 1816

October brought a return to normal temperature levels for the first extended period since April. Afternoon maximums climbed consistently into the 60s and 70s; the highest of the month at Middlebury reached 76°. Indian summer conditions prevailed with warm hazy afternoons and relatively cool nights. The most conspicuous feature of the weather in the early autumn was the smoky atmosphere caused by forest fires across northern New England, New York, and Ontario. The *Northern Centinel* at Burlington reported that smoke drifting from extensive burnings along the shore of Lake Champlain from Plattsburgh to Ticonderoga became thick enough to impede navigation on the lake, and to reduce visibility along western Vermont highways to about two rods.

This additional weather hazard of the season continued until after mid-October when the first general precipitation in many weeks arrived. Snow fell across northern sections, and rain turned to snow in the south. Haverhill, New Hampshire, in the northern Connecticut River Valley, reported 12 inches on the ground on the morning of the 18th, and at Williamstown, Massachusetts, the ground was white. Presumably, the entire state received a white blanket at this early date.

The precipitation quenched many of the forest fires temporarily, but some rekindled and continued until the first extended wet spell since the spring arrived. Rain fell at Williamstown every day beginning October 21 and ending October 28; the seven-day total reached 2.56 inches, ending the agricultural drought and extinguishing the forest fires for good.

Remainder of 1816

Many thought the early-season snowfall portended the coming of severe winter conditions, but the wind flow changed at the end of October, and a delightful late autumn and early winter followed. Middlebury's thermometer rose to 72° and Williamstown's to 71° on November 4. Fields greened up again, permitting cattle and sheep to graze well into December. Rainfall continued sparse; the September to November total at Williamstown amounted to 6.14 inches, or 77 percent of normal for the three months. So Vermonters were given a welcome respite during late autumn and early winter from the adverse conditions that had persisted for six long months.

Severe conditions did not descend until January 17th, but then in a spectacular fashion with heavy thunder and lightning and a unique display of St. Elmo's fire during a snowstorm on the higher ground of Windsor and Windham counties.

Agricultural Effects

The editor of the Windsor *Vermont Journal* surveyed the current state of agricultural affairs in the first issue of October:

> The Weather—Never perhaps in this vicinity appeared more gloomy and cheerless than at present. It is extremely cold for the time of year, and the drought [sic] was never before so severe. We have had several frosts in this county, and we believe in every county in the state, in every month during the last fourteen mos. The late frosts have entirely killed the corn. It is not probable that enough will get ripe for seed for next year. There is not sufficient hay to winter the cattle upon, and nothing with which to fatten them this fall. The crops of english grain [wheat] were generally good, but nearly everything else has failed, and even this can hardly be converted into flower [sic] for want of sufficient water to keep mills in operation. In short we are something like the soldier, who "had no allowance, and no kettle to cook it in."
>
> *Vermont Journal*, Windsor (Oct. 7, 1816).

No contemporary reports have been located in the press or in personal diaries concerning actual famine across northern New England. After the passage of time, however, such stories crept into local histories with constant retellings enlarging on the woes of the season. Many households certainly were forced to improvise their eating fare as a result of the failure of the corn crop. In the upper Connecticut River Valley the following winter and spring of 1817 became known as the "mackerel year," since fish from the Atlantic seaboard were transported to the interior communities and eagerly purchased by those whose regular supply of pork had been cut off by the scarcity of corn feed for their swine.

The Swanton area of northwestern Vermont became a trading center for people from the hill country towns to the east who came to trade their maple sugar products for fish caught along the Missisquoi River.

99

At least ten large seines were set up in the river to increase the normal catch. Maple products were considered legal tender during the winter of 1816-17.

People thanked their Scotch neighbors for having introduced oatmeal to the American table. The mill at Ryegate had to run full time during the winter to produce a sufficient supply to enable many to survive the shortages of the season. Wheat and small grains generally replaced corn as the principal bread staple.

The almost complete failure of the corn crop posed a serious problem for farmers, not only for the supply of breadstuff and animal feed, but also for the seed for next year's crop. The price of corn seed soared as the full impact of this disastrous season became apparent. Corn in the New York market normally varied from 75¢ to $1.12 a bushel, but the value reached $1.35 after the August frosts and eventually climbed to $1.78 by January.

One Newbury resident went south to Connecticut and brought a flat-boat up the Connecticut River loaded with corn, which he sold for $2.50 a bushel (the normal price in the area previously had been 50¢ a bushel). To balance this unfavorable aspect of one Vermonter's character, there was the oft-told story of Squire Thomas Bellows of Bellows Falls who sold limited quantities of seed corn at normal prices to anyone coming down from northern Vermont. His act of compassion and charity was later celebrated in a popular poem.

At a meeting of the Philadelphia Society for Promoting Agriculture on October 30, 1816, a resolution directed a special committee to determine the effect of the unusual spring and summer weather on the crops of the nation.* A circular appeared in many newspapers requesting agriculturists to communicate their experiences, and a selection of the replies was published in the annual report of the society. While dispelling the idea that there had been a general crop failure, the summary emphasized that Indian corn and hay were much below normal in yield as a result of frosts and drought, that certain interior sections suffered much more severely than coastal and southern localities, and that many individual farm economies were disrupted by the adversities of the season.

Only two of the reports dealt with northern New England communities. From Bowdoin College at Brunswick, Maine, Professor Parker Cleaveland reported: "good crops of wheat and rye were raised in Maine; but the corn has been almost entirely cut off." Another account from Colonel Samuel Morey of Orford, New Hampshire, across the river from Fairlee, complained of the unusual shortness of the season since the killing frost on August 21 had cut down "almost the entire corn crop."

*Philadelphia Society for Promoting Agriculture, *Memoirs*, Vol. 4, 1818.

THE VERMONT WEATHER BOOK

The Philadelphia Society for Promoting Agriculture summarized general conditions and added a favorable aspect of the season: "Vegetation was very materially affected by the weather. The small grains in abundance, and very good, but the crops of hay were deficient, and Indian corn, by the frosts of August was almost lost. But for the inclemency of the weather the inhabitants were compensated with a greater share of health than had ever been known since the settlement of the town."

Social Aftermath of the Summer of 1816

The psychological impact of the discouragements of the successive cold seasons must have been great on many industrious farmers who saw their hard labors go for little or no return. The late-starting winter of 1816-17 turned extremely severe in February and March to add to the discomfort, and then another backward spring in 1817 once again culminated in light June snows and frosts with the temperature falling almost as low as the year before. Many farmers wondered if the New England climate had not changed permanently for the worse, and whether living prospects might not be brighter elsewhere.

Coincidentally, the vast Ohio Country was opening for settlement now that the Indian menace had been removed by the outcome of the War of 1812. A "Genesee Fever" (referring to the Genesee River Valley in the vicinity of Rochester, New York) had prevailed in the late 1790s and early 1800s when many Vermont families picked up stakes and moved to the reputedly rich soils of western New York State. In the summer of 1817, a similar "Ohio Fever" infected many of the discontented in the Green Mountain country with an irresistible urge to move west and seek their fortunes in a different clime.

A number of factors must have influenced the hard decision to leave farm, family, and friends. Professor Lewis D. Stilwell, the historian of emigration from Vermont, wrote: "Perhaps the heaviest blow which hit Vermont in any of those years was the "cold season" of 1816...Something, it seemed, had gone permanently wrong with the weather; and when this cold season piled itself on top of all the preceding afflictions, a good many Vermonters were ready to quit." He mentioned the adversities arising from the economic prosperity and collapse attending the War of 1812, the tragic epidemic of "spotted fever" in 1813 and 1814, and the gradual wearing out of the soils after years of intensive cultivation. Furthermore, the growing population forced younger sons to find new farms for themselves and their families. Most looked to the West.

101

Atmospheric Circulation during the Summer of 1816

A simulation of the upper-air circulation over North America for the three summer months of 1816 would place a ridge of high barometric pressure extending from the Hudson Bay region of central Canada southward over the lower Great Lakes and western New York State, then continuing along the crest of the central and southern Appalachian Mountains. This atmospheric feature blocked the normal west-to-east movement of airstreams crossing the continent along the lines of latitude. Forced to the north, the airstreams followed a sinuous wave pattern around the periphery of the obstacle posed by the high-pressure ridge. Downstream from the ridge to the east lay a trough of low barometric pressure, stretching from Labrador and Newfoundland southward over the ocean to the vicinity of Bermuda.

Airstreams from the Pacific Ocean rounded the northern extension of the high-pressure ridge over central Canada and plunged as cool, dry airstreams southeast into the trough of low pressure off the Atlantic coast. Successive surges of air that had acquired the characteristics of the Canadian subarctic entered the northeastern corner of the United States from May through September. The persistence of the airflow from the northwest effectively prevented tropical air from the Gulf of Mexico and low latitudes of the North Atlantic from reaching New England.

Whether conditions in the source region of polar air around Hudson Bay were colder than normal, we have no direct measurements, but reports appeared in Montreal and Quebec newspapers of lakes still frozen in interior Quebec in mid-June and of heavy snows falling along Hudson Strait at the end of the month.

Occasional frosty conditions during July and August are not unknown in the Green Mountain region during the present century, but the combination of five successive days of frost in early June, a continuance of cool nights in July, the twin visits of freezing temperatures in late August, and the climatic late September black frosts have not been reported in any season since 1816.

UNUSUAL WEATHER EVENTS, EXTREMES, AND EXPERIENCES

UNUSUAL WEATHER EVENTS, EXTREMES, AND EXPERIENCES

> "There is a sumptuous variety about New England weather that compels the stranger's admiration—and regret. The weather is always doing something there; always attending strictly to business; always getting up new designs and trying them on the people to see how they will go."
>
> Mark Twain. "Speech on the Weather" at the New England Society's seventy-first annual dinner, New York City, December 22, 1876.

Many unusual events have occurred in the skies over Vermont, and people are affected in rather extreme ways by untoward atmospheric events they experience. The following section presents several of these events as described by Vermonters in town histories, weather records, personal diaries, and newspapers to illustrate the "sumptuous variety" in the realm of Vermont's atmosphere.

"Darkness at Noon"—The Total Solar Eclipse in 1806

The greatest total eclipse of the sun since the settlement of the United States by Europeans occurred on June 16, 1806, and the southern portion of Vermont lay within the path of totality. The shadow of the moon raced on a east-northeast track from the Pacific coast of southern California to the Atlantic coast of Massachusetts and New Hampshire.

A path of the moon's shadow (line of totality) has crossed the coterminous United States on only one occasion since then: in June 1918 when it ran from the state of Washington diagonally southeast to the peninsula of Florida.

The celebrated eclipse of 1806 occurred near noontime and close to the summer solstice, conditions very favorable for a long duration of totality. At this season the moon is closest to the earth, a position called perigee, so the diameter of the moon's shadow is wider than at other times of the year. The maximum extent is about 163 miles. The duration of the 1806 eclipse in New England was about four minutes and fifty seconds, close to the maximum possible for this latitude.

The center of totality crossed Massachusetts from west to east on a line passing through North Pittsfield-South Deerfield-Wachusetts Mountain near Princeton-Bedford-Salem. Totality took place at Salem from 11:25:26 to 11:30:14, a duration of four minutes and 48 seconds. At noon the center of totality was at 42°N and 66°W, or about 275 miles east of the tip of Cape Cod.

The northern limit of totality encompassed all Bennington and Windham counties along with a small slice of southern Rutland County and about one-quarter of southern Windsor County. The line ran slightly east-northeast from the New York border at West Pawlet, through South Wallingford and Ludlow, to Ascutney on the Connecticut River. The shadow raced eastward at a speed of about 1,300 miles per hour.

No scientific observations of the eclipse were known to have been carried on within Vermont. The best source of local information appeared in the pages of the *Vermont Gazette, an Epitome of the World*, published at Bennington by Benjamin Smead, who was presumably the editor and took an intense personal interest in the event. In the issue of June 16, he devoted space to a general description of the coming eclipse based on "Darkness at Noon," a current Boston pamphlet describing the coming event. In the next issue of June 23, the editor presented a description of his own observations at Bennington, and in the issue of June 30, he was able to provide accounts of the eclipse as viewed at Albany, New York, along with brief excerpts from the Rutland and Windsor newspapers.

Editor Smead's description of the solar spectacle follows:

> The singular event of a total eclipse of the sun, on Monday, the 16th instant, has excited much philosophical investigation among the people of this vicinity. Many enquiries have arisen in the minds of the observers—Where was the center of the shadow?—How far did it extend to the north and south?—And what was the period of the exclusion of the sun-beams at the different places between the northern and southern extremities of total darkness? The day was fair; not a cloud appeared here above the horizon—a few minutes before the total shade, many stars appeared—a considerable change of atmosphere from heat to cold succeeded—a general darkness was instantaneous—the fowls hastened to their nightly shelters—the cold increased rapidly—the dew fell as copiously in 4 or 5 minutes as in the preceding night—stars appeared all around the horizon—and in Bennington the total shade continued about 3 m 50 s as was calculated by several watches present. The scene was sublime beyond description, and occasioned serious reflections. On the return to light, it was peculiarly animating. The effulgence of the heavens burst in upon us sudden as the lightning, and

displayed a glory peculiarly striking—It gave some idea of that scene, when *the morning stars sang together, and all the sons of God shouted for joy.* It dispelled the gloom as sudden as the conception of the human mind.—Day beamed forth again—the air was soon restored to its mild temperature—the birds began to carol their morning notes—the stars retired from the view of mortals—and pleasure filled every countenance.

We learn that the eclipse was total in Pawlet, about 38 miles north of Bennington; and that it was not so at Wallingford, about 45 miles north of this place. It was conjectured generally, that the center of the shade was about 20 miles south of us, and much farther north in its whole course than was calculated by astronomers. We wait with solicitious expectations to hear further on the subject of this singular and curious phenomenon.

Vermont Gazette (Bennington), June 23, 1806.

Liberty Pole Enables the Rescue of Many in 1811 Flood on the Poultney River

Mr. Corbin's family, and some children (13 in number) were in a house in the midst of this (now) extensive river. The waters rising and roaring on each side of them, and the remaining banks which sustained the house were momentarily giving way. The women and children were shrieking for help, but for awhile no mortal could devise any means for their relief. Behold the scene! Mr. Corbin, on the shore, beheld his wife, children and aged mother, in this deplorable situation; and the parents of some school children, also beheld their tender offspring on the brink of a watery grave. At length, Mrs. Corbin gave up all hopes of surviving the calamity, shut her doors, and concluded that she and her family, &c. must in a few moments go into the eternal world, and was committing herself and all to God. But in this awful moment the liberty-pole* was thought of, and instantly brought, which was long enough to reach the bank on which the house stood, and by means of this, a number of men passed over, and relieved the family and the other children. Within 15 minutes after they were relieved, the house went down stream. Mr. Orria Cleveland lost his life in attempting to save this distressed family; he was found the same day, but such was the violence of the waters, that not the vestige of a garment was left on him; he was buried the next day.

Reverend Lemuel Haynes, Middletown, Vt., July 24, in *Vermont Gazette* (Bennington), July 30, 1811.

*Liberty poles were erected during the first decade of the 1800s on the central greens of many towns and villages by Democratic Societies that supported Thomas Jefferson and the philosophical basis of the French Revolution.

The Luminous Snowstorm in January 1817

In the evening of the 17th of January 1817, as I was returning from a neighbour's house in company with another young man, between the hours of ten and eleven, we observed that the snow was falling very fast., and that there were frequent flashes of lightning. The singularity of the last-mentioned appearance, at that season of the year, particularly attracted our attention. After noticing it for some minutes, we passed on, and when at the distance of a few rods, we observed an appearance of light or fire upon the tops of the stakes in the fence and bushes by the side of the way. The novelty of these appearances rendered them, in our opinion, worthy of particular attention. And the probability of their soon disappearing made us more anxious that there might be others to witness the same fact. We accordingly called the people who were in a neighbouring house; but instead of the lights disappearing, they seemed to shine with additional brightness, and to appear where they were unobserved before; on our hats, on our hair, when our heads were uncovered, and on a woollen mitten, when held above our heads. These lights appeared on all substances, which extended to any considerable distance above the surface of the ground, and approached to a point; but they had not all the same form. For the most part the light appeared like a spark or star, but in some instances it resembled a blaze, in form of an inverted cone of about two inches in height, and three fourths of an inch in diameter at the base. These blazes emitted a hissing sound resembling that of the water in a tea-kettle just before it boils, which could be heard distinctly at the distance of six or eight feet. After viewing them for some time, we proceeded homewards, and in passing over ground somewhat higher, found this appearance to increase, so that our hats and shoulders were almost covered with the lights; and when we spit, the small particles of saliva, when at a little distance from the mouth, assumed a shining appearance. After going about fifty or sixty rods, we came to the side of a lot of standing timber, where these appearances were not to be seen, except on the tops of some high apple trees on the other side of the road. The falling of the snow, and the height prevented our ascertaining whether there were lights on the tops of the forest trees. We did not leave the standing timber, until we had come down on low ground, where we saw no lights until we got home. About twelve o'clock, we returned to the place where we first saw the lights. The snow was yet falling very fast, and there appeared to be still more of the before-mentioned lights; but we saw no blazes.

Where these lights were seen, was a tongue of elevated land, extending from the north, down between two quite small branches of Williams' river. The whole length of way we travelled that evening was about a mile. Where we first saw the lights, was very near the top of this ridge of land, descending steeply to the east, for about a hundred rods to a small stream, not sufficient to turn a mill, and to the west, with a more gradual descent, for a mile to another stream a little larger. There are no very considerable streams in this vicinity.

The wind was not strong, but a light breeze blowing from the northeast, and the weather was not so cold as usual for that season of the year. The flashes of lightning, when we first observed them, which was a little past ten o'clock, were very frequent four or five in a minute; but the claps of thunder were seldom, and these appeared to be at a distance, and not heavy. But the flashes of lightning soon became less frequent, more vivid and more usually attended with thunder. About twelve o'clock, there was a number of very sharp flashes of lightning, attended with quite heavy thunder.

The above-mentioned blazes of light were precisely like the appearance upon a wire, in a dark room, overcharged with positive electricity. All that I can say of the circumstances necessary to produce these blazes is, that we observed them on stakes in the fence higher than other substances near them, and extending to a considerable depth in the ground, and when they were removed, the blazes disappeared, and were succeeded by stars.

> Joel Manning, jun., of Union College, then residing at Andover, Vermont, in John Farrar, "An account of a singular electrical phenomenon, observed during a snow storm accompanied with thunder," *Memoirs of the American Academy of Arts and Sciences* (Boston, Mass.), 1821.

Fatal Thunderstorm at Brattleboro in 1818

On the 15th Inst. we had a terrible storm of wind and rain attended with flashes of lightning and peals of thunder that might appall the stoutest heart. Several buildings were struck by lightning, but with little damage except a large barn of Lemuel Ball, filled with the products of his industry the past season, which was consumed to ashes. Two lads were milking in the barn: his son and a relative who lived with him. – about 16 years of age, who with the cow he was milking was struck dead. Mr. Ball's son, though struck down was not so stunned but that he effected his escape before the flames seized him and carried the awful news of his companion to his father. Ball flew to the barn, but the flames enveloped it, so fierce, he could not approach. The body was not rescued until almost destroyed. The charred remains were interred on the 17th when an affecting discourse was delivered to a large audience by Rev. Hollis Sampson from Job V. 6 to 9 inclusive.

The Brattleboro Reformer, October 27, 1818, in Abby Hemenway, *Vermont Historical Magazine*, 3-3, p. 10.

The "Extraordinary Darkness" of November 1819

This phenomenon first attracted my attention on the morning of the 9th of November 1819. I rose at a quarter before seven, and found it much darker than it ordinarily is in the evening at the time of full moon. It snowed fast for about an hour; this was succeeded by a moderate rain, which continued most of the day. Being occupied, I took no further notice of the uncommon darkness till about nine o'clock. At this time, the obscurity, instead of diminishing, had considerably increased. The thermometer stood at 34°. A strong, steady, but not violent wind blew from the south.

The darkness was so great, that a person, when sitting by a window, could not see to read a book, in small type, without inconvenience. Several of the students in the college studied the whole day by candlelight. A number of mechanics in this village were unable to carry on their work without the assistance of lamps.

The sky exhibited a pale yellowish-white aspect, which, in some degree, resembled the evening twilight a few moments before it disappears. Indeed we had little else but twilight through the day; and such, too, as takes place when the sun is five or six degrees below the horizon. The color of objects was very remarkable. Every thing I beheld wore a dull, smoky, melancholy appearance. The paper, on which I was writing, had the same yellowish-white hue as the heavens. The fowls showed a peculiar restlessness that was remarked in them during the total eclipse of the sun in 1806. Some of then retired to roost. The cocks crowed several hours incessantly, as they do at the dawning of the day.

At 3 P.M. the sky brightened up a little, but in the evening the darkness became more extraordinary. A person could not discern his hand, held directly before his eyes. It was next to impossible for a person to find his way even in streets where he had been long accustomed to walk.

The sun was concealed from our view, nearly the whole time, from Monday evening to Friday morning. It did occasionally appear, but was always of a deep blood-red colour; and the apparent magnitude was at least one third larger than usual. This was very striking on Friday, about nine in the morning. A dense, yellow low vapour was then passing slowly over its enlarged disc. The spectacle was viewed by many with astonishment. . .

The darkness was observed throughout the northern portion of this state, and in several parts of Canada. At Montpelier, about forty miles northeast of this place, it is said to have been greater than it was here. A gentleman, from that town, informed me that the darkness there was so great, that the Speaker of the House of Representatives could not distinguish the countenances of the members so as to determine who was addressing him. The same gentleman added, that where he stopped to dine, he was obliged to make use of a candle to distinguish the different kinds of food which were placed before him.

In the small quantity of water, which fell from the atmosphere, I did not observe any extraordinary colour, or smell, or taste. It is stated in *Le Courier du Bas-Canada*, "that the water was of a black colour, as if it had been impregnated with a large proportion of soot; and that several persons, who had tasted it, discovered the taste of soot. This colour the water retained a considerable time." I have read remarks of a similar kind in the newspapers from various parts of New England. Had the fall of water been more copious, I should probably have noticed the peculiarity above described.

The appearance of the heavens during the late period of darkness was very much like that which is frequently occasioned by extensive fires in the woods. An effect, similar in kind, but far inferior in degree, was produced a few years since by the fires, which raged several weeks, and consumed most of the underwood on the Green mountains opposite this place. The darkness observed at that time was very considerable, and the sky was of a pale yellowish-red aspect.

> Frederick Hall, "On the extraordinary darkness that was observed in some parts of the United States and Canada, in the month of Nov. 1819," *Memoirs of the American Academy of Arts and Sciences.* (Frederick Hall was Professor of Mathematics and Natural Philosophy in Middlebury College, Middlebury, Vermont.)

The Tale of the Frozen Woman— December 1821

It is our painful duty to record one of the most distressing incidents which has ever occurred in this vicinity. Early on Wednesday morning last [Dec. 19], Mr. Harrison G. Blake, an inhabitant of this town [Cambridge, Washington Co., N.Y.], left home with his wife and one of his children, about 15 months old, intending to visit his father-in-law and other friends beyond the Green mountains. They reached Arlington, Vt. in safety, about 11 o'clock, and soon in the afternoon proceeded on the road leading over the mountain through Sunderland. As they ascended they found the snow much deeper than they had expected, and after two or three miles no sleigh had passed since the late snow, and no path or track was to be found. With much labor, however, they slowly pressed forward breaking their road through snow more than two feet deep, on the side of a steep, rugged mountain and nearly five miles away from any human habitation. Here night overtook them, and to augment their dismay, their horse fatigued by such protracted exer-

111

tions in the snow, began to lag and at length stopped. After some deliberation, they concluded to loose him from the sleigh and make another vigorous effort to save themselves and their child. The following extract from a letter dated Stratton, Dec. 21, written in Mr. Blake's name by his attending physician, to his friends here, exhibits all the additional particulars of this melancholy affair which have come to our knowledge.

"My wife rode, and carried the babe a short distance only, when she said she could ride no further. She then alighted and told me she would walk as far as she could after me, and answer to my calls. I took her mantle and gave in return my great coat and mittens. Her responses soon became so low that I could not hear them, nor could she probably hear my calls for help, or my addresses to her. She advanced but a short distance, before she left our dear babe, wrapt in my great coat in the snow. She did not travel more than 150 rods when she became so chilled and frozen that she sank, never to rise again. She was found alive next day, but survived only a few moments. I was about 40 rods from her in advance, obviously in a perishing condition. But a few more hours, Dear Sir, and I, too, must have been beyond the power of human assistance. I called aloud repeatedly before I became benumbed with the cold; but all to no avail. We were all providentially found yesterday afternoon, and carried to the nearest habitation in this town. Our babe was found half a mile from my deceased consort with his face naked and in the snow; it smiled affectionately when taken up; it is not frozen except one foot, and that not badly. My feet are both frozen half way to my ankles, my hands are also much frozen, and today indescribably painful."

THE VERMONT WEATHER BOOK

A true and particular statement of the sufferings of Harrison G. Blake and the death of his wife on the Green Mountains taken from the relation of the above mentioned Mr. B. and from Mr. Richardson, the young man who first found the sufferers.

Washington Register (Cambridge, N.Y.), Dec. 27, 1821.

The Torrent of July 1830

Probably the most dramatic incident in all Vermont flood history occurred on the New Haven River at New Haven West Mills in Addison County on July 26, 1830. Several families were surrounded by the rising waters with tragic results. The story was given wide publicity in the newspapers of the day and found a permanent place in Vermont literature through the publication of a pamphlet by one of the survivors.

The reader has now seen us in three parties successively afloat on the relentless surges. After the escape of Mr. Willson and son, nineteen individuals were left to the enraged elements; and all now had left the spot where they were at first surrounded, which was now completely overflowed.

I can speak of none now, but those who were on the same float with myself. After leaving the barn and urging our way into the middle of the stream, as abovementioned, but few minutes elapsed before we arrived at the narrows. The first swell we rode in safety. But our little float was rifted in our attempt to rise the second, and in a breath of time, none were above the surface of the water. This was a solemn parting. With my son, it was my last. As we neared the monstrous swells, my son in a tone of suppressed emotion observed to me—"in a few minutes we must all be in eternity," a broken reply in confirmation of the assertion was all I could, we spoke no more!

On our being plunged in the stream, I strove to keep at a distance from the surface, and to this expedient I probably owe my life; and, as my son was an expert swimmer, he most likely strove to keep on the surface, and by consequence, perished among the floating timbers.

Having used all my energy to keep below the surface, and being borne swiftly by the current, I passed through the rapids, a distance of 50 or 60 rods without sustaining any injury or receiving any water in my lungs. Heavy lumber, however, as I had passed under the last swell, plunging in almost a perpendicular direction, I received several severe blows upon my head and different parts of my body, by the force of which I was driven to an enormous depth in an eddy where the water stood at thirty or forty feet. At this moment I was driven in contact with one of those who commenced their journey with me on the raft. As my arm passed about the body

113

· THE ·

TORRENT;

OR

AN ACCOUNT OF A DELUGE OCCASIONED BY AN UN-

PARALLELED RISE OF THE NEW-HAVEN RIV-

ER, IN WHICH NINETEEN PERSONS WERE

SWEPT AWAY, FIVE OF WHOM ONLY

ESCAPED, JULY 26TH, 1830.

BY LEMUEL B. ELDRIDGE.

—◆◇◆—

"There's a divinity that shapes our ends,
" Rough-hew them how we will."—*Tragedy of Hamlet.*

Middlebury:
PRINTED AT THE OFFICE OF THE FREE PRESS,
BY EDWARD D. BARBER.

1831.

Not every Vermont storm, freshet, or "deluge" results in a book, but a severe rainstorm on July 26, 1830, flooded the New Haven River, causing 14 deaths and many thousands of dollars of property damage. Lemuel B. Eldridge, whose son was lost in the flood and who himself was seriously injured, wrote an account of "The Torrent" in book form which was printed in Middlebury in 1831. Nearly a century later the great flood of 1927 became the subject of many illustrated books and pamphlets.

I discovered no motion of life; I perceived by evident marks upon his clothing that it was no other than the corpse of my son! Cold indeed was the embrace! But no more cold than short. He was immediately swept from me, and I labored for the surface.

I gained it at length, and after receiving air sufficient to prolong life, was again dashed beneath, and again arose. I then secured

THE VERMONT WEATHER BOOK

a plank floating near, and by this means kept myself most of the time above. After floating thus about three quarters of a mile from the place where I embarked, I caught a hill of corn on the river bank, and remained till day break in three or four feet of water defended by unremitting exertion from the drift-wood which I distinguished only by the dark spot it occasioned on the white field of foam. By the grey of the morning I discovered but few rods above me, upon a ledge on which I had designed to land, two of young men who had been wafted thither upon a large timber, which the water, although it over flowed it, had not sufficient force to drive past the ledge. About the same distance below me Mr. Farr, (who could not swim) having been wafted to the place by means of two planks which he seized on falling from the raft, had lodged upon a rolling stump and forced its prongs in the earth. He had but just gained the position, when the older of the two young Stewarts floated by within speaking distance, not daring to trust an attempt to land, he passed the ridge on which we were secured and immediately again came in deep water. Floating a distance of about sixty rods, immediately above the cluster of rocky islands and a rapid in Otter Creek, to have passed which would have been death inevitable, he caught the top of a young tree, and climbing above the reach of the waves, remained until helped ashore next morning by his friends.

Mr. Farr and myself, after having been assisted to the ledge by the younger Stewart brother and young Farr, by means of a series of poles, confined by strings rent from their garments, and thrust to us, were, with the young men, brought ashore upon a raft constructed for the purpose by those assembled on the bank of the stream which still continued swollen and violent.

Five of the nineteen we have seen safe; the remaining fourteen are now quietly slumbering, alike regardless of the calls of friends, and the tumult of the waters.

Lemuel B. Eldridge, *"The Torrent: or an account of the deluge occasioned by an unparalleled rise of the New-Haven River, in which nineteen persons were swept away, five of whom only escaped, July 26, 1830."*

When Stars Fell on Vermont—The Great Meteor Shower of November 13, 1833

The Leonid meteor shower annually puts on the best spectacle on the 12th and 13th of November when the orbit of the meteors streams across that of the earth. Every thirty-three years an exceptional display occurs with thousands of shooting stars per hour streaking across the sky. The best in the nineteenth century appeared in the early morning hours of

November 13, 1833, when it was observed from the Atlantic to the Pacific coasts. Much folklore arose from the event and was given classic form in Carl Carmer's book, *Stars Fell on Alabama*.

At Randolph

Nov. 13 This morning rose at half past five; weather cold & windy— meteors falling in all directions & of all sizes; at any moment from that time to day light several might be seen at the same time—as day light advanced the atmosphere became so cloudy that they could not be seen, though till the clouds obstructed the sight, they appeared to be snowing with unabated rapidity.

William Nutting, manuscript diary, Vermont Historical Society, Montpelier, Vermont.

At Newbury

The meteoric shower of November 13, 1833, was one of the most wonderful sights ever witnessed. The night was perfectly clear, and about ten o'clock the display began. Thousands of meteors fell, some of them with dazzling brilliancy. The flashing was incessant, many at the same time falling in all directions. Some were awakened from sleep by the glare, and the superstitious thought that the end of the world had come."

History of Newbury, Vermont, p. 265.

Rebuke to Snow and Cold Weather in May 1834

Thou vile perfidious Winter King
What meanest thou that thou should bring
Thy storms of snow and cold in May?
Where smiling spring should hold its sway
Has thou not had enough of sport
In March and April as thou ought
To let alone at this late day
The ruddy face of blooming May!
Thou ought it indeed to be ashamed
The fare of spring thus much to maim
This shameless villanous to see
Such work dishonest made by thee
Now if you would good manners show
Stop now and send us no more snow
Recall old Boreas to his cell

And let mild Zephyr's *houres* tell?
Brush now your gloomy clouds away
And let bight Sol illum the day
And cause chilled Nature to revive
And grass and budding foliage thrive
Retire now to the polar star
To Boreas regions north afar
Nor dare again thy frost intrude
Those six months in our latitude
For if you do snowy Heclor throw
Its burning lava round below
Arrest they imprudent career
And blot thy kingdom from the year

James Johns, Huntington,
May 1834

"The Most Remarkable Exhibition" of Aurora Borealis in 1837

But the most remarkable exhibition of this meteor, which has fallen under our own observation, was in the evening of the 25th of January, 1837. It first attracted our attention at about half past 6 o'clock in the evening. It then consisted of an arch of faint red light extending from the northwest and terminating nearly in the east, and crossing the meridian 15 or 20° north of the zenith. The arch soon assumed a bright red hue and gradually moved toward the south. To the northward of it, the sky was nearly black, in which but few stars could be seen. Next to the red belt was a belt of white light, and beyond this in that direction, the sky was much darker than usual, but no clouds were any where to be seen. The red belt, increasing in width and brightness, advanced toward the south and was in the zenith of Burlington about 7 o'clock. The light was then equal to the full moon, and the snow and every other object from which it was reflected, was deeply tinged with a red or bloody hue. Between the red and white belts, were frequently exhibited streams of beautiful yellow light, and to the northward of the red light, were frequently seen delicate streams of blue and white curiously alternating and blending with each other. The most prominent and remarkable belt was of blood-red color, and was continually varying in width and intensity. At eight o'clock, the meteor, though still brilliant, had lost most of its unusual properties.

Zadock Thompson, *History of Vermont* (1842).

117

On the Snow Storm of 11th of June 1842

Old Boreas sounds! What are you doing
  Are you crazy gone and mad.
To set it now in June a snowing
  By the hoky 'tis too bad.
Not content Jack Frost to send us
  Would you spoil our summer quite
With snow storm come on us tremendous
  Putting Nature's smiles to flight
You are a villian thus to come on
  With your wintry host like this
O may the sun his forces summon
  And routing make your snow flakes hiss
What sense is there in all this mocking
  Of the farmer's hopes and pains
The step you've took is really shocking
  Tis rascally I'll say again
To get you back unto the pole and
  There remain till the fit time
And ne'er again intrude your cold in
  June for it will never suit this clime
Let summer's bland and gentle zephyrs
  Hold alone their genial sway
Touch not the corn without twere tougher
  It is no business for thee
Go get thee home to Arctic ocean
  There expend your snow and cold
Let it no more make a motion
  This way till October's told
May your keeper overtake you
  Drag you back to your dark cell
  And better manner'd may he make you
  Guard you till you're rational
Shame on the freak you have committed
  Look round, then blush and hide your face
See that it is to June unfitted
  Sneak home and better keep in place.

          James Johns, Huntington, June 11, 1842

Snowballs in the Air—1854

At Royalton, Vt., during a recent snow storm, while it was snowing fast, and the wind high, a white appearance in the air resembling a flock of white geese was observed, but it was soon ascer-

tained to be balls of snow collected in the air by a thousand whirlwinds. They increased in size and number for about twenty minutes, when they fell to the ground covering something more than an acre. On examining the spot where they fell, balls were found from one to twenty inches in diameter.

The Daily Free Press (Burlington), Jan. 20, 1854.

"A Ride in a Tornado"—November 1869

We alluded yesterday to the fact that our townsman Dr. Carpenter, while on a professional errand to Huntington, was caught in the terrible tornado of Saturday. He gives a graphic description of his adventures. Coming up a hill he encountered the first gust of the storm at the crest, and found the air full of shingles, tree-tops, fence-rails, and such like "motes." Going down thence to cross a stream, the bridge swayed so fearfully he dared not cross, until there came a lull in the tempest. The next bridge was impassable, its roof lying in the meadow, and the body being tipped over. Guided by motion from a man on the opposite bank—for not a syllable could be heard in the roar of the storm—he went down the river bank some distance before finding a place to cross. After reaching the road again he met two of the selectmen going in a wagon to examine the fallen bridge, and watched them for a few rods till coming out from under a bank. The storm struck them and instantly overturned the wagon, men and all. Presently he met a man and a woman in a wagon; they got on well enough for awhile, till the man incautiously rose in his seat to look ahead, when the wind hustled man and buffalo-robe over the dashboard on to the horse's back! At another place a flock of sheep had gathered under the lee of a fence, the best shelter they had, and were safe for a time, but presently a stronger gust swept fence, sheep and all right up the hill, and whirled them promiscuously from the crest.

Entering the piece of wood where the doctor was in greatest peril, at first all was quiet, but as he progressed, decayed trees began to fall, and as he went on to the further side, hemlocks of a foot and a half through were twisted off by the gale like saplings. One such had fallen right across the road; and while the doctor was deliberating whether to cross the gutter on foot or in his wagon, an elm tree top swept through the air behind him, and striking the ground, its six-inch branches were splintered up like ovenwood. The doctor "stood not upon the order of his going" after that!

But, despite the perils of such a ride, the doctor reached his patient, whose house alone was left, and that stood under the spire of the church, which swayed as if about to fall any moment. It is possible the doctor made his visit a short one!

Passing one house, he saw a man come out with a huge basket

119

in his hands. The wind caught it, but the man held it, till his involuntary strides grew too long for comfort, when he let go, and away went the basket like a balloon. Many a funny incident he saw, although of many the fun at the time overshadowed by the absolute danger to life and limb.

The Free Press and Times, morning edition (Burlington), Nov. 23, 1869.

The "Yellow Day" on September 6, 1881

At Burlington

After a very sultry night, the sun got up very late Tuesday morning, and for some time after the time for sunrise it was very uncertain whether we were to have daylight or not. When it came, slowly and strugglingly, it was not like ordinary daylight, but was a sort of unearthly, artificial, reddish brown glimmering, changing as the hours went on to a yellowish glare. The peculiar light modified the hues of almost everything. The artificial light in the houses burned blue by contrast with the deeper yellow of the surrounding atmosphere. The freshest complexions had a sallow tinge. The colors of the flowers were altered, the purples taking a reddish tint, the blues changing to green, and the green of the landscape having a strange unnatural shade. The entire sky was canopied by a dull, yellowish curtain of mist. The air was free from smell of smoke or sulphur. What it all meant—whether it was the tail of a comet, or a volcanic eruption of Camel's Hump, or owing to the President's removal or the excursion of the Odd Fellows or the beginning of the fulfillment of Mother Shipton's prophecy, nobody could tell. But everybody agreed that it was very remarkable, and that they never saw anything like it. Towards eleven the peculiar appearance wore away somewhat, though the sun remained partially obscured and could be looked at with the naked eye, as if through smoked glass, at noonday. In the afternoon the sun was like a crimson wafer against the sky, and passed out of sight about 5 o'clock behind the thick haze. The explanation of it all, doubtless, is a veil of smoke high up in the air. While the wind near the surface of the earth yesterday was from the south, the upper currents were from the west—and these, no doubt, brought over this region, a stratum of smoke from the western forest fires, too high up to affect the senses in the ordinary way, which veiled the sun and colored the landscape after this unusual fashion. Towards sundown last night the smell of smoke was distinctly perceptible in the air.

The Daily Free Press and Times (Burlington), Sept. 7, 1881.

At St. Albans

The sky presented a very singular appearance early in the morning. The atmosphere was smoky, as it had been for several days, and so continued through the day. At four o'clock, and from that time until a little after six, the dense haze which filled the heavens was red as the redness of the evening sky. The color changed to a phosphorescent green, which lasted till about half-past seven. One peculiarity of the luminous haze was that it cast no shadows. It was as light underneath the trees as it was above them, and the landscape was as devoid of perspective as a Chinese painting. Everything had a weird and ghostly appearance. The gas flames were white as electric light, and it was not until the sun had risen above the horizon that the natural appearance of things had been restored.

St. Albans Messenger, Sept. 6, 1881.

Severe Cold Possible Cause of Vermont's Worst Railroad Wreck in 1887

Soon after leaving White River Junction, the Montreal Express from Boston met disaster about 2:30 A.M. on February 5, 1887, while crossing the bridge over the White River at Hartford. The engine and front cars were halfway across the 610-foot long bridge when a rear car derailed and caused the coupling with the four rear cars to break. After bumping over the ties for some distance, these fell to the river ice fifty feet below and soon caught fire, reducing to ashes the cars and those unfortunates trapped within. The superstructure of part of the bridge also burned.

The investigation traced the accident to a split rail near the approach to the bridge. Though the cause of the rail failure was never fully determined, it was thought at the time that the intense cold prevailing might have been responsible for the fracture of the rail. The temperature fell as low as $-5°F$ that night in the vicinity. Of the 90 passengers believed to have been aboard, the death toll amounted to 34 persons and the injured list to 49.

First Biennial Report of the Board of Railroad Commissioners of the State of Vermont, December 1, 1886, to June 30, 1888.

"Twas the Montreal Express" begins the Vermont folksong that tells the story of the famous Hartford wreck of February 5, 1887. Part way across the 610-foot bridge that crosses the White River north of Hartford the rear car derailed and caused the coupling with the four rear cars to break and send the cars fifty feet to the ice below. Fires started from the coals in the heating stoves in the cars, and many who had survived the fall from the bridge died in the fire. This tragedy in which thirty-four died and forty-nine were injured helped lead to the outlawing of stoves in railroad passenger cars.

THE VERMONT WEATHER BOOK

Ball Lightning at Burlington, or UFO?

Ball lightning is a relatively rare form of lightning, consisting of a reddish, luminous ball, of the order of one foot in diameter, which may move rapidly along solid objects or remain floating in mid-air. Hissing noises may emanate from such balls and they sometimes explode noisily, but may also disappear without any sound. The physical nature of ball lightning is one of the minor mysteries of atmospheric electricity. Following is one person's detailed observations of this phenomenon, as manifested in downtown Burlington.

I was standing on the corner of Church and College streets just in front of the Howard Bank and facing east, engaged in conversation with Ex-Governor Woodbury and Mr. A.A. Buell, when, without the slightest indication or warning, we were startled by what sounded like a most unusual and terrific explosion, evidently very near by. Raising my eyes and looking eastward along College street, I observed a torpedo-shaped body some 300 feet away, stationary in appearance and suspended in the air about 50 feet above the tops of the buildings. In size, it was about 6 feet long by 8 inches in diameter, the shell or cover having a dark appearance, with here and there tongues of fire issuing from spots on the surface resembling red-hot unburnished copper. Altho stationary when first noticed this object soon began to move, rather slowly, and disappeared over Dolan Brothers' store, southward. As it moved, the covering seemed rupturing in places and thru these the intensely red flames issued. My first impression was that it was some explosive shot from the upper portion of the Hall furniture store. When first seen it was surrounded by a halo of dim light, some 20 feet in diameter. There was no odor that I am aware of perceptible after the disappearance of the phenomenon, nor was there any damage done so far as known to me. Altho the sky was entirely clear overhead, there was an angry-looking cumulo-nimbus cloud approaching from the northwest; otherwise there was absolutely nothing to lead us to expect anything so remarkable. And, strange to say, altho the downpour of rain following this phenomenon, perhaps twenty minutes later, lasted at least half an hour, there was no indication of any other flash of lightning or sound of thunder.

Four weeks have past since the occurrence of this event, but the picture of that scene and the terrific concussion caused by it are vividly before me, while the crashing sound still rings in my ears. I hope I may never hear or see a similar phenomenon, at least at such close range.

Bishop John S. Michaud, quoted in William H. Alexander, "A possible Case of Ball Lightning," *Monthly Weather Review* (Washington) (July 1906): 325-27.

123

A Perilous Ride in the Flood of 1927 at Barre

Perhaps the most trying physical experience of any survivor of the flood was that which befel Helge Carlson, one of the two men who endeavored to rescue the Thomas boys at Webster avenue and who was swept down the flood for a distance of nearly half a mile, as stated in the previous chapter. Few could go through the same experience, and live – and Carlson himself declares: "I guess I was pretty lucky." Lucky he was, and plucky, too.

Carlson is a well set up young man 25 years of age and of Swedish birth. He came from Sweden five years ago, and, as he says, he was in the water a good share of the time over in the old country; so he is a good swimmer. He needed to be that night.

It was only by chance that Carlson came to be involved in the tragedy at Webster avenue and, consequently in his own great experience. He was on his way to visit a countryman in the south end of the city Thursday night and came opposite just as the people were crying for help. Carlson volunteered to be one to man the boat that had been brought to effect a rescue, if possible. He joined McNulty, previously mentioned. They went out in the boat carrying a roll of slack rope. But let Carlson tell the story!

"The current was so swift it carried the boat rapidly down. We grabbed a house as the boat was swept along. Pulling hard we got the boat up to the house. We asked 'who's goin' first?' and they said the kids. So the kids got into the boat and we started to pull the boat along a rope that was hitched between the shore and the house. The current was so strong the boat tipped over. I didn't see the other fellow after that and don't know how he got out. I grabbed a kid who had been sitting in the front of the boat near me and lost hold of him when the boat went over. Pretty soon I heard a gurgling noise and swam over toward it and found a kid – I don't know whether it was the same one or not.

"I tried to swim with the kid but the current was so strong I couldn't hold him, and the boy slipped down and grabbed hold of my leg. Gee! I can feel him there now. But he lost his hold and I didn't see him after that. I swam to a pole that was sticking out of the water, and there I sat. Guess I stayed there about half an hour, when along came some big object that knocked pole and me into the flood. The river was full of floating pieces and I got hold of a raft and climbed on top.

"We rushed along...the flood was full of things which kept bumping into my raft but I kept hold. Pretty soon the water began to act kind of funny, you know, and all at once me and my raft dived right over a waterfall and we went bottom side up (he had gone over the Trow and Holden dam). I lost hold of my raft when it went out from under me. I came up to the top of the water and logs and boards kept banging into me."

The distance from the Trow and Holden dam to the Prospect street bridge is not more than 500 feet but the water was very swift that night, and Carlson was right in the midst of the swiftest part of the current. He was hurled along, keeping his head above water by paddling although greatly weakened by his tremendous exertion in staying on his raft from Webster avenue.

"When I saw there was a bridge ahead of me, I made up my mind here was my chance. So just as the flood swept me up to it I lunged and grasped hold of the underpart. I must have hung there ten seconds when the current knocked my hands off and away I went. I don't know how I got under the bridge, nor the railroad bridge just below – perhaps I went over this bridge. Then I came to the stonesheds there and as the water swept me toward them I made another lunge. This time I was lucky; I got a hold of a compressor pipe that sticks out the end of the Hoyt and Milne shed. I hung on for dear life and then when I had got strength enough back I pulled myself up the pipe and then got to the top of the Hoyt and Milne shed.

"There I stayed all night – gee! it was cold. Along about 8 o'clock the next morning (Friday) someone came along and found me on the roof of the shed and they got me home. I was pretty weak for a long time. Yes, I guess I was pretty lucky," was the way the young fellow ended his modestly related story of his perilous night ride on the bosom of the most disastrous flood that ever swept this valley.

Dean H. Perry in Barre In The Great Flood of 1927

The Total Eclipse of the Sun on August 31, 1932

The first total eclipse of the sun to be viewed directly over Vermont since 1806 occurred early on the afternoon of August 31, 1932, when the northeast corner of the state witnessed a short two minutes and twenty seconds of totality. The peak of the Vermont phase of the eclipse occurred between 2:15 and 2:18 P.M., EST.

The central line of the shadow of the sun entered the state from Canada near Derby Line, Orleans County, passed southeast almost over Island Pond, Essex County, and left the state at the Connecticut River in Guildhall, the shiretown of Essex County. The length of the traverse of the 115-mile wide center of totality in Vermont was about 40 miles. The southern edge of the solar blackout was marked by a line running southeast from the Canadian border near Alburg on Lake Champlain

125

to Fairlee in Orange County on the Connecticut River. St. Albans, Montpelier, and Chelsea lay just outside the southern edge of totality.

Weather conditions over Vermont were generally unfavorable for the professional astronomers since a veil of cirrus clouds at high elevations, along with some middle clouds of the alto-cumulus type prevented a clear view of the beautiful corona and other phenomena attending an eclipse. The general public, however, witnessed the almost complete darkening of the atmosphere and sensed the general eerie feeling present.

At Island Pond

Island Pond's eclipse period practically was a failure for astronomers located here. Eager crowds lined hill tops, fields and highways, some with cameras, but only ten percent located in this territory were able to get a thorough view of the eclipse at all times. Some sections, however, gave several local people a view of the corona and the beautiful circlet around the dark outline of the moon. Frequent spots of clear sky in a few sections here gave some of the partial phases of the eclipse....

The moon's shadow was a magnificent sight here. Frogs were heard croaking, pigeons went to roost, birds were bewildered, cows came to barns, a flock of wild ducks on the lake here seemed lost, darting back and forth in the water, then suddenly rushing to shore to hide in the bushes. Two local people were overcome with fright at the totality, necessitating physicians.

At Morrisville

About 1,000 persons gathered in Morrisville and on Elmore Mountain to observe the eclipse of the sun today.

Dr. Lloyd Robinson, Morrisville dentist, reported that a temperature test he made during the phenomenon showed that the thermometer dropped from 92 degrees to 69½ degrees within a period of approximately 35 seconds. The lowest temperature was recorded during totality, after which the mercury again rose to around 92 degrees.

The street lights in Hyde Park were plainly visible from Elmore Mountain, a distance of approximately seven miles.

At Danville and St. Johnsbury

Diamond Hill in Danville, about 2,000 feet above sea level, was the mecca for 250 observers of the eclipse, many from other states than Vermont being in the interested group. After the first 15 minutes the clouds passed away and a fine view was obtained. There were many in parts of St. Johnsbury disappointed as a dense cloud obscured the sun as well as the moon during totality, though on Harris Hill and some other spots the sky was fairly clear and the corona and Bailey's Beads plainly visible.

The above accounts of the solar eclipse at several locations in Vermont are from the *Burlington Free Press*, Sept. 1, 1932.

126

Snow Rollers at Burlington—March 1936

Some very excellent "snow rollers" were formed on the lawn of the Burlington, Vt., Weather Bureau station and in the immediate vicinity on the morning of March 13, 1936....

Conditions for the formation of snow rollers must be perfect: a level or gently sloping, smooth surface, a light, fluffy snow of proper density, temperature slightly above the freezing point, and a fresh wind. All of these conditions were present.

On the evening of March 12 the ground was bare of snow, except for a trace in patches. Rain fell lightly until midnight, then moderately to 12:30 A.M., .08 in. being recorded in this half-hour period. Sleet began at 2:25 A.M., changing to a moist, fluffy snow at 2:45 A.M. Snow continued until 6:30 A.M., total depth 2.8 inches.

The temperature was 34° at 2:00 A.M., 33° at 3:00 A.M., 32° from 4:00 A.M. to 8:00 A.M., and 33° at 9:00 A.M. The wind was northwest between 2 and 3:00 A.M., south 3-4:00 A.M., and southwest, 4-5 and 5-6 A.M. The velocity had been light, about 9 miles per hour during this period.

Shortly after 6:00 A.M., the wind shifted to south and increased in velocity. Seventeen miles were recorded from 6-7 A.M., 19 miles from 8-9 A.M., and 24 miles from 8-9 A.M. The max velocity reached was 27 miles from the south at 8:32 A.M.

Snow rollers began to form about 6:30 A.M. They continued to "roll" until about 9:00 A.M. when probably the increasing temperature and bright sunshine, which caused surface thawing, prevented their further occurrence.

Some of the rollers were nearly perfect in shape. They varied in size from very small to about 13 inches in diameter and about 18 inches in width. Literally hundreds of them were in view over the adjacent fields, the smaller ones being much more common. The phenomenon had never been observed by any member of the office nor "by the oldest inhabitant."

The theory is advanced that possibly the sleet may have played an important part in the formation of the rollers. An examination of a number of rollers seemed to show an icy lining or "frosting" between the several layers.

M. W. Dow, Weather Bureau, Burlington, Vt., in *Bulletin of the American Meteorological Society*, 63 (February 1937).

Polluted Hailstones at Bennington in 1950

On June 27, 1950, a hailstorm struck Bennington, causing considerable controversy because of the metallic nuclei about which many of the stones had formed. A plausible explanation was that bits of slag entered a strong uprising current in the thunderstorm when it passed low over the high stacks of a coke company in Troy, N.Y. about 25 miles to the west. These particles, some of which contained coal, after several accretions of ice, reached a diameter of 1¼ to 2 inches, causing minor damage to windows, automobiles and gardens. The main street was covered with a white carpet.

Climatological Data, New England, June 1950.

The Great Windstorm of November 1950

Addison County was slowly recovering Thursday from the effects of the worst disaster in its history. Damage exceeding one million dollars was caused last Saturday night when a hurricane swept up the Champlain Valley. The area between Pittsford and Vergennes was devastated almost beyond description. Hundreds of trees were uprooted, miles of fences ruined, seven out of every ten houses suffered roof damage, some slight, some severe. Barns were blown down, 1,000 head of cattle are dead, families are temporarily homeless.

Electric power is gradually being restored to the stricken area and telephone service is increasing daily. As of Thursday, Middlebury, Brandon, Salisbury, New Haven and Vergennes were cut off from the outside world, except for emergency messages handled by radio.

Labor and roofing materials were scarce. One bright spot was the favorable weather conditions. If snow and cold will hold off for a few more days, many repair jobs will be accomplished and suffering decreased....

The Vermont Extension service conducted a survey Thursday,.... Their check discloses $840,000 damage to farms. Added to this is the $150,000 damage to Middlebury College buildings, damage to scores of homes and village properties as well as losses in homes due to wind and water.

Their investigation disclosed the following information: Estimated number of farm houses and tenant houses totally or partially damaged, 50; estimate of total loss in connection with these buildings, $30,000; estimated number of dairy barns totally destroyed or partially damaged, 35; estimated loss in connection with these buildings, $300,000; dairy animals killed in the county, 1,000; estimated value, $200,000; loss of machinery, including barn, stable and field equipment, $50,000; loss of miscellaneous buildings, small barns and hen houses, $250,000; total damage to fruit orchards, $10,000.

Addison Independent (Middlebury), Dec. 1, 1950.

The storm which struck the northeastern United States with devastating force last Saturday, was the most destructive of its type

128

in the history of Middlebury College. Preliminary estimates of damage to college property are placed at $100,000.

Despite the violence of the storm and the extensive damage which resulted, no serious injury to a Middlebury student, member of the faculty or administration was reported.

The Memorial Field House, Pearsons Hall, Mead Chapel, and the Student Union Building sustained heaviest damage from winds of gale force and driving rains. No college building escaped damage completely. Slates and screens were blown off all dormitories, and windows were smashed as the 75 mile per hour winds battered the buildings. Extensive loss came from rain, which poured through damaged roofs, soaking walls, ceilings, floors, and flooded basements after seeping through as many as three stories. Falling trees knocked out communications and electric power on the campus. Tree losses were heaviest on the lower campus, where nearly 20 trees were felled by the winds.

Mead Chapel suffered extensive damage to windows, its spire, and roof. Two solid cement urns were torn from their bases on the railing of the steeple and crashed through the Chapel's slate roof. One of the urns fell on through the ceiling of the balcony, smashing a pew. Windows on the south wall of the building were broken and torn from their frames. The circular window on the pediment was blown in. The basement was flooded, but students bailed it out on Sunday to prevent damage to an electric motor and choir robes.

Most of the damage in Pearsons Hall came from rain which poured through gaping holes in the roof tops and seeped through partitions and floors. Occupants on the rooms on the east side were evacuated with their belongings to the hall or other dormitories on Sunday when it became apparent that ceilings were in danger of collapsing. The work of repairing and replastering damaged rooms was immediately begun. It is indefinite when the work will be complete and students can move back in.

Addison Independent (Middlebury), Dec. 8, 1950.

Snow Rollers at Morrisville in February 1973

Light snow had fallen intermittently for several hours in southwesterly winds ahead of an arctic front approaching from the west. Air temperatures were 28-32°F. By 2300, 4 February, about 3" of very light fluffy snow had accumulated over a hard snow base. No snow was falling at this time and eye-witnesses (David Mudgett et al.), who observed the initial formation of the rollers in a large yard under bright lights, report that strong gusty winds began to

Once in a while the combination of light, sticky snow and high winds creates nature's snowballs. This was the case throughout the Lamoille River Valley from Bakersfield to Stowe on February 4, 1973. Strong, gusty winds scoop out clumps of snow which are then rolled along the surface. Here Kelly Moodie of North Wolcott kneels beside a large snowball in a photograph taken by the late Robert Hagerman.

dig into the light snow cover, compressing and scooping out clumps which were tossed into the air and rolled along the surface.

Snow conditions were light, wet, and sticky so that the clumps gathered snow as they rolled. About this time or perhaps a bit later, the front moved through and winds shifted to northwesterly and continued gusty. The clumps apparently continued to roll, forming cylindrical shapes with dimensions about 12" in diameter and lengths about 18". In the clear, cold air following the front (approx. 25°), the roller formation ceased as the winds slackened – or at least became less gusty. The soft core in many rollers apparently blew out leaving a donut-like hole.

An area encompassing the Lamoille River Valley from Bakersfield to Stowe in northern Vermont was covered with rollers by morning, and they persisted for several days. Tracks behind the rollers were erratic, but were generally SW to NE or NW to SE. Rollers developed in sheltered as well as open spaces.

E. B. Buxton, *Weatherwise*, 26-2 (April 1973), 87.

HISTORIC STORMS OF VERMONT

TROPICAL STORMS AND HURRICANES

Although it lies far from southern latitudes and without an exposed seacoast, the Green Mountain region has experienced the impact of storms originating in the tropics, ranging from a severe hurricane such as bisected the state in September 1938 to the remnants of the circulation of an erstwhile tropical storm wandering up the Connecticut River Valley attended by copious showers. Fortunately, visitations of mature hurricanes retaining their destructive wind force have been few.

In first rank, one must place the mini-hurricane of August 1788, the Great September Gale of 1815, and the New England Hurricane of September 1938. A grade below would be the September Hurricane of 1821, the October 1846 Storm, the Tropical Deluges of July 1850, the Central New England Hurricane of 1878, the August Storm of 1893, Hurricane Carol in 1954, Hurricane Hazel in October 1954, and Hurricane Donna in 1960. In the past decade, there have been three brushes with storms that classified as hurricanes before reaching Vermont latitudes in diminished form: Belle in August 1976, and David and Frederic in September 1979.

The main effect of tropical storms on Vermont is the enhancement of the normal rainfall. The intruders from the south usually bring northward vast quantities of warm, humid, tropical air that results in heavy precipitation when the cyclonic whirl reaches New England. The deluges form an important part of the rainfall budget of Vermont, helping to maintain a satisfactory water table level in the late summer and early fall when precipitation from other sources is often scanty.

When excessive rains fall from a tropical storm, or when a previous wet period is augmented by tropical storm rainfall, too much water may

133

collect for the streams to handle, creating a flood. The high water attending the Hurricane of 1938 provided an example of what can happen when tropical-born rains fall on a soil saturated from previous storms, and the Vermont Flood in 1927 resulted when the remnants of a tropical storm drifted over the region and loosed vast quantities of precipitation on a sodden soil. More recently, ex-hurricane Belle in August 1976, though deprived of its damaging winds before landfall, caused very heavy deluges in the Green Mountains that brought about soil erosion and losses to agriculture.

The Mini-Hurricane of August 1788

On August 19, 1788, an intense hurricane with a small central core and a tight circulation cut a swath across Vermont from southwest to northeast, with the center passing over Bennington and Windham counties and through Windsor County during the early afternoon hours. Destructive winds continued for less than thirty minutes, but enormous forest destruction occurred. A child was killed near Dummerston and a man near Plainfield, New Hampshire.

The *Vermont Gazette* reported after the blow: "The northern post informs us that a great deal of damage has been done in the different towns of his rout [sic], to the corn &c., and many cattle have been killed by the falling of trees during the late heavy storms."

Nathan Perkins, a Connecticut missionary, visited the towns in the Champlain Valley next spring, and observed: "Almost half of ye trees in ye woods down by ye violence of ye wind last year."

The storm was still remembered in the middle of the nineteenth century when Abby Maria Hemenway was gathering local material for her monumental *Vermont Historical Magazine*. The contributor to that record from Dummerston wrote:

> When the storm arose, dense black clouds rolled up from the north-west; the tempest winds roared with fearful sounds of gathering power; lightnings flashed vividly through the moist atmosphere; the thunder deepening and crashing as if it would rend a world; then came the violent rain and the rushing hurricane with one full blast that swept whole forests to the ground. No swaying of trees, back and forth, but one continued rush of the mighty wind prostrated every tree in its range for miles up the West River valley, and along the west side of the high range of hills in the central part of the town. Black Mountain was left bare of its vast forest of large trees. Many cattle were killed, buildings unroofed, and one little child lost its life....
> People were greatly frightened during the tempest, and many went into the cellars for fear their houses would be blown down.

Several men, the next day, took their axes and butcher-knifes and went over the fields, and killed what cattle were living that were injured beyond recovery.

Abby Maria Hemenway, *Vermont Historical Magazine*, vol. 5, 119-20.

The Great September Gale in 1815

A full-fledged hurricane from the tropical seas bisected New England during the daylight hours of September 23, 1815. From an entry point on the shore of Long Island near Saybrook, Connecticut, the center moved almost due north, passing west of Worcester and into New Hampshire in the vicinity of Peterborough. Next morning it was reported near Quebec City, though in a much diminished form.

The only contemporary reports for Vermont appeared in the press. The *Brattleboro Reporter* told of heavy rains on Friday night and Saturday morning, September 22 and 23, and a resultant flood. The *Vermont Journal* at Windsor mentioned the "violent storm on Saturday last," but gave no details. Nothing appeared in the *Vermont Gazette*, published at Bennington on the west side of the mountains. Since all of Vermont lay to the west of the storm track, wind flow would have been from the northeast at the start, shifting through northwest to southwest as the center of the hurricane drew abreast and passed on to more northerly latitudes.

Wind speeds normally would have been much less in Vermont than in the eastern sector over New Hampshire and western Maine where enormous structural and forest damage occurred, but rainfall in the Green Mountains would have been heavier than farther east.

The September Storm in 1821

A powerful Atlantic hurricane smashed into the Norfolk, Virginia, area about noon on September 3, 1821, and then raced along the Maryland, Delaware, and New Jersey coastlines to make a second landfall on extreme western Long Island over Jamaica Bay. It has been known as the Norfolk-Long Island Hurricane. Although of lesser physical stature than the Great September Gale of 1815, it did considerable damage in the vicinity of New York City and western Connecticut. Noah Webster, then living at Amherst, Massachusetts, described "a violent hurricane from SE commencing in evening. It prostrated trees and overset some sheds and houses. It was violent at Amherst and as far north as Brattleborough." Farther north the editor of the *Vermont Republican & American Yeoman* at Windsor commented: "In this vicinity the wind did not blow so violently, as there described; but the rain fell in great abundance, and the roads were much damaged."

The course of the storm north of the Massachusetts line cannot be traced. The Windsor report would indicate that the main effect in that vicinity came from heavy rains, and this is confirmed by observers at Bennington and at Williamstown, Massachusetts. The Williams College observer, Chester Dewey, noted "a violent rain in the evening" of the 3d and the wind at 2:00 and 9:00 P.M. observations was out of the southeast. He measured 2.97 inches of rain during the two days of the storm. At Bennington, Benjamin Harwood thought the weather "threatening" and "very gloomy," but mentioned no unusual wind in his diary. It would appear that the storm lost its tropical structure in passing over the Berkshires of Connecticut, although it retained its rain-making capabilities well into central New England.

Tropical Storm of July 1850

A tropical storm reaching northern latitudes in July is a rarity. One did so on July 19-20, 1850, and the entire Middle Atlantic and New England areas experienced unexpected heavy deluges that raised streams to high flood levels. The path of the storm passed over Chesapeake Bay and Delaware Bay with heavy rains falling on Washington, Baltimore, and Philadelphia, west of the storm track.

Since all stations in New England experienced a southeast wind flow, the center of the storm must have passed up the Hudson River Valley or a few miles to the west. Nevertheless, the rainfall in western New England proved excessive; probably the raised terrain of the Berkshires and Green Mountains produced the necessary lifting of the tropical airstreams. Burlington reported 3.23 inches, and Montpelier 5 inches.

The *Vermont Chronicle* at Windsor summarized the flood situation: "In this section of the country, it appears that North and East the flood was somewhat less than in 1830. South and Southwest it was much heavier, and thus was therefore, taken altogether, the greatest storm and flood that ever occurred here within the memory of man."

Three Tropical Storms in the Season of 1893

August 24, 1893 — A tropical storm with considerable destructive power was located near New York City early on the 24th. Twelve hours later it was astride the New Hampshire-Maine border east of Berlin, New

136

Hampshire, having crossed the extreme southeast portion of Vermont. Extensive damage to trees and small buildings was reported on the eastern side of the Green Mountains. Rainfall at Northfield amounted to 0.99 inch and winds did not attain gale force.

August 29, 1893 – A second tropical storm within a week hit Vermont while moving on an unusual track. It turned east over Lake Ontario and traveled at a fast clip to reach the Quebec-Maine border by the evening of the 29th. The center did not cross northwest Vermont, but the accompanying winds gave that section a sweep of its destructive force. Northfield registered a maximum of 27 miles per hour from the west and received a heavy rainfall of 2.11 inches. No doubt, winds were higher and rainfall heavier in Grand Isle and Franklin counties. Extensive forest and structural damage was reported in the press.

October 13-14, 1893 – A third storm from the south to affect Vermont moved on a track similar to Hazel in 1954 and occurred on about the same dates. From a position in central Virginia on the evening of the 13th, it advanced to Ontario in twenty-four hours. A very low barometer reading of 28.74 inches was reached at Toronto. At Northfield a wind of 47 miles per hour from the south was clocked along with a rainfall of 0.45 inch.

The New England Hurricane of 1938

The impact of the New England Hurricane in September 1938 in Vermont stands second only to that of the devastating Flood of November 1927 in the disaster history of the Green Mountain area. In total scope, the hurricane exceeded the flood since both high and low ground were affected, while the losses in 1927 were confined mainly to the stream and river valleys. The devastating wind force in 1938 reached to the very tops of the ridges and crests of the mountains, and great destruction took place in the remote forest areas of the state. The flood waters equaled or exceeded those of 1927 and 1936 on the smaller streams in the southern counties.

One hundred and fifty years had passed since the center of a vigorous tropical storm packing hurricane-force winds had cut a swath within the Green Mountains, and only a few antiquarians were aware of the destruction attending that visitation in 1788. Hurricanes were thought to be the concern of New England's seashore residents, and, even there it had been so long since a mature hurricane had paid a visit that people were not hurricane conscious and were completely unprepared for the tragic events that unfolded on the afternoon of September 21, 1938.

Along with the rest of New England, Vermont had experienced a rainy period since the middle of the month that raised streams to near flood

137

The Hurricane of 1938 takes its place alongside the flood of 1927 as a major, twentieth century, Vermont weather catastrophe. While Vermont shared the big blow with the states to the south, its forests suffered most severely. Crossing the state from Brattleboro to Burlington, the big winds felled hundreds of thousands of trees including a great many Vermont sugarbushes.

stage, but no apprehension was expressed. The weather forecasts for Wednesday, the 21st, called for continued cloudy conditions with some rain. Officials in Washington released an advisory concerning a tropical storm east of the Bahama Islands that was supposed to follow a normal course by recurving off the coast of the Carolinas and head northeast well offshore into the North Atlantic shipping lanes.

Instead, the hurricane center sped directly north-northeast, accelerating its forward pace to more than 60 miles per hour and bursting on the Long Island shore about 2:45 P.M. on the 21st before warnings of an apparent change in direction toward the mainland could be distributed to the weather bureau stations in New England.

The central eye of the storm roared across Long Island Sound, entered Connecticut just west of New Haven, and then dashed north along the west flank of the Connecticut Valley, close to the line of present Highway 10, to the vicinity of Northampton and Brattleboro. There the center veered to the northwest, crossing the Green Mountains on a diagonal about along a Marlboro-Weston-Rutland-Brandon-Middlebury-Vergennes line. A low pressure of 28.68 inches was registered at Burlington at 8:00 P.M. when the center was over the lake to the west. Other pressure readings at Vermont stations were Rutland 28.74 inches, Northfield 28.77 inches, and Newport 28.90 inches.

Wind speeds in the three southern New England states attained full hurricane force of 73 miles per hour or more at ground level. By the time the center entered Vermont, the friction of passing over irregular land surfaces and the cutoff from the source of warm, oceanic moisture were already diminishing the wind power of the storm.

Atop Mt. Washington in northern New Hampshire, well away from the storm track, gusts reached 162 miles per hour and speeds of 136 miles per hour were sustained for a full minute. The Hurricane of 1938 raged at an average of 118 miles per hour for a full hour between 5:00 and 6:00 P.M. when the center was about to enter Vermont. At the surface in Vermont, both Burlington and Northfield recorded a maximum sustained wind of 47 miles per hour. Amherst, Massachusetts, reported 58 miles per hour, and Hanover, New Hampshire, 46 miles per hour. These were sheltered valley locations; speeds on the higher mountain slopes and ridges were probably 50 percent higher.

The structure of a hurricane is divided into sectors. The eastern side, and particularly the northeast quadrant, is considered within the "dangerous semicircle" of the storm. Wind speeds are higher than in the western side, being augmented by two factors: the pressure gradient determining the wind force is usually tighter in advance of the center, as indicated by the rate of fall of the barometer; and the forward momentum of the system, in this case between 50 and 60 miles per hour, augments the total wind force.

Under normal conditions, wind speeds east of the diagonal from Brat-

The Hurricane of 1938 did millions of dollars of damage to the state's forests, but it also took its toll in cities and villages. This car parked on Church Street in Barre was simply in the wrong place at the wrong time. Photo courtesy of Aldrich Public Library.

tleboro to Burlington should have been higher by a factor of two to one over those to the west. Forest destruction was reported to be much greater in the Connecticut River Valley and on the eastern slopes of the Green Mountains than on the western slopes and in the Champlain Valley.

The wind devastation on a farm in Cabot in northeastern Washington County may be considered typical of what happened to areas that experienced the eastern sector of the Hurricane of 1938:

> The farm of William Walker of Cabot was located in the path of this big wind. Mr. Walker tapped 3500 trees in 1937; this destructive wind uprooted or broke off 2400 of these trees so that he could tap only 900 in 1939. Mr. Walker operated two orchards, one was entirely destroyed, the other which was near the house was uprooted and therefor required a lot of work clearing roads before he could make sugar again.
>
> Another maple orchard that was nearly destroyed belonged to Peyton Walbridge on the Walden road. He had 5000 maple trees with a retail market for all of his product; when the storm was over he had only 1000 trees left.

Harold P. White took this view, showing the demolished bridge and street in Wilmington. Water stood six feet deep in the street when the cement bridge collapsed. From The Flood and Hurricane of 1938, *published by* The Brattleboro Daily Reformer.

That meant that this family's livelihood was nearly destroyed overnight for one faces the real tragedy of this when they realize that it takes from forty to sixty years to grow a maple large enough to produce maple sap abundantly and most orchards are 100 years or more old.

As the worn out pastures of Vermont become timber lands and hurricanes continue their unpredictable destruction each fall, I believe it would add confidence to timber land owners if a deductible wind insurance became available to the coming generations of Vermonters.

If the eastern side of an advancing hurricane is known as the windy sector, the western side is the wet sector since the heaviest rainfall usually falls to the left of the line of advance of the storm center. The strong upper-air currents from the southeast attending the approach of a hurricane are forced to rise over the cool wedge of air brought into the surface circulation by northwest and north winds. In rising, the southeast airstream is cooled and much of its moisture condenses into raindrops. The Hurricane of 1938 behaved in this fashion in southwestern New England. Another factor came into play from western Massachusetts northward. The steadily rising terrain gave additional lift to the moist airstreams and wrung even more rainfall from the clouds.

All stations in Vermont, except in the extreme northeast at a distance

141

from the storm's path, received 2.75 inches or more of rainfall during the final twenty-four hours of the extended rain period. Wilmington near the southern border reported the most with 4.45 inches, and Newport on the Canadian border had the least with 1.19 inches. The addition of these excessive amounts to what had already fallen during the previous four days raised stream levels to high flood stage. Details of the floods created by the Hurricane of 1938 are found in the chapter devoted to floods.

The Non-Hurricane Windstorm of November 1950

The vast storm striking the northeastern states on November 25, 1950, caused damage throughout Vermont second only to that of the New England Hurricane of 1938. Although winds probably reached in excess of the hurricane criterion of 73 miles per hour, the storm was technically not an authentic hurricane since it originated over the southern Appalachian Mountains and not over a tropical sea and did not have the warm core structure characteristic of true hurricanes. With the path leading northward along the crest of the Appalachians, both flanks suffered severe lashings by high winds and excessive precipitation, heavy snow falling to the west and heavy rain to the east. Although Vermont lay in the warm sector well to the east of the track of the storm center over central Pennsylvania and western New York State, winds of gale force swept all of the Green Mountain region, with the western section suffering the greatest damage. The vast disturbance has been named the Great Appalachian Storm of November 1950 as a result of the widespread destruction from the Carolinas north into Canada.

The western valleys and the northwestern agricultural plain of Vermont suffered most for two reasons: they were closer to the very deep center of the storm where wind speeds were highest, and the southeast flow induced by low pressure to the west crossed the crest of the Green Mountains and descended to the lower elevations of the Valley of Vermont and the Champlain Valley. The downslope motion created a dynamic fall or gravity wind condition that increased the total wind force sweeping the rolling countryside.

Hurricane Carol in August 1954

Hurricane Carol followed about the same path and time schedule of the Great September Gale of 1815. It smashed over the eastern end of

Long Island, came ashore close to the mouth of the Connecticut River near Saybrook, Connecticut, and then raced almost due north, passing west of Worcester, Massachusetts, about noon, and through the entire length of New Hampshire during the afternoon. Carol was of lesser stature than the Great September Gale of 1815 and the Hurricane of 1938, but it provided a lively reminder of the destructive potential of a mature tropical storm.

Vermont lay entirely in the western sector of the storm where wind power was lower (the highest wind at Burlington reached 38 miles an hour from the north). Total losses in the state were estimated at only $300,000, compared to an overall loss of $438 million in Long Island and New England.

The heavy rainfalls normal in the western sector of the hurricane circulation developed over Vermont; but with no antecedent rain period, serious flooding did not arise. The heaviest amounts occurred over eastern Vermont close to the storm's path through New Hampshire: Wardsboro 3.49, Cavendish 3.11, and Woodstock 2.73 inches. Bennington in the southwest reported 2 inches, and St. Albans in the northwest, 2.24 inches.

Hurricane Edna in September 1954

Hurricane Edna followed only twelve days after Carol, but pursued a path farther east that missed most of the mainland area of New England. From a landfall on Martha's Vineyard shortly after 1:00 P.M. on September 11, 1954, Edna crossed Massachusetts Bay and the Gulf of Maine, trending northeast, to a second landfall on the Maine coast near Eastport.

Vermont lay well to the west and suffered only minor inconveniences. Burlington's highest wind was 37 miles per hour from the north on the 11th. Mt. Washington, much nearer to the storm track, raised a wind force of 112 miles per hour from the west.

Rainfall throughout the Connecticut River Valley generally ran between 1 and 2 inches. Lemington in the north reported 2.23 inches, Bennington in the southwest 1.25 inches, and Burlington in the northwest, 0.95 inch.

Hurricane Hazel in October 1954

The third hurricane of the active 1954 season, Hurricane Hazel, passed well to the west of Vermont over central and western New York State on the afternoon and evening of October 15. On this occasion Vermont lay in the eastern sector of the hurricane and experienced relatively high winds. Burlington's peak of 70 miles per hour just missed the hurricane classification, and Mt. Washington clocked a lower speed than during Carol, but equal to Edna's 112 miles per hour. Rainfall ranged from over an inch in southern sections to zero in the extreme north. Wardsboro measured 1.43 inches, but the rain gauge at St. Albans remained dry. The damage in all New England totaled only $35,000, despite the high winds in Vermont.

Ex-Hurricane Connie in August 1955

Ex-hurricane Connie moved across central Pennsylvania and extreme western New York State during the daylight hours of August 13. Although it had lost much of its wind punch by the time it reached Vermont, Connie continued as a rain producer of the first order. The southerly airstreams in its eastern sector, brought north from the tropical central Atlantic, produced tremendous rainfalls when forced to rise over the hills and uplands of New England.

Thunderstorms occurred on the 11th throughout Vermont in the tropical air that had overlain the area for several days. The heaviest amount was 4.09 inches at Reading Hill, but most amounts were below 1 inch. The hurricane circulation with clouds and rain arrived on the 12th and produced twenty-four-hour amounts as high as 4.04 inches at Readsboro on the 13th. The three-day totals attained 8 inches at three stations: Whitingham 8.29 inches, Readsboro 8.08 inches, Weston 8 inches, and there were several other stations over 6 inches. The catch tapered off to the north: Newport reported only 0.79 inch attributable to Connie, and Canaan in the extreme northeast, only 0.39 inch.

Ex-Hurricane Diane in August 1955

Within a week of Ex-Hurricane Connie another tropical storm came north, but along a different path. From a landfall in North Carolina, Ex-Hurricane Diane's path curved over Virginia, southeast Pennsylvania, and central New Jersey before heading east-northeast along the southern shore of Long Island during the daylight hours of August 19. This path enabled Diane to develop a strong upper-air flow from the south and to carry vast amounts of moisture from the warm ocean surface, which were deposited on the land of southern and central New England in record amounts. In the hills of western Massachusetts near the Connecticut border, Westfield caught the storm's highest amount, an amazing fall of 18.15 inches in twenty-four hours. The canopy of heavy rainfall just reached into southern Vermont. Vernon reported 2.95 inches, Mays Mills

144

3.45 inches, Reading Hill 2.57 inches, and Whitingham 2.15 inches. But some central and northern Vermont stations received no precipitation at all on the 18th and 19th.

Hurricane Donna in September 1960

The most recent tropical storm to raise full hurricane winds in New England struck during the late afternoon and evening of September 12, 1960, when Hurricane Donna moved over eastern Long Island about 3:00 P.M. and then traveled from Connecticut to Maine on a New London-Framingham-Durham-Rumford-Fort Kent line. Donna was a large, but rather loosely organized storm. Its passage over land through eastern New England caused its eye, or relatively calm center, to become distorted, yet it maintained very low barometric pressures as far west as eastern New York and western Vermont. Pressure fell below 29.00 inches over all Vermont except for a segment northwest of a line from Rutland to Newport.

The highest surface wind at Burlington reached 34 miles per hour from the northwest. Atop Mt. Washington the anemometer spun at 112 miles per hour during a peak gust from the west-northwest.

Heavy rainfall covered the entire Green Mountain region during the 11th and 12th, with a number of stations reporting in excess of 5 inches. Somerset topped all with 6.33 inches. In the southwest, Bennington had 4.49 inches, in the northwest Burlington measured 3.04 inches, and in the northeast Newport received 3.07 inches.

Tropical Storm Doria in August 1971

Tropical storm Doria cut a diagonal across New England from southwest to northeast during the morning and early afternoon hours of August 28, 1971. From New York City, it moved near Northampton, Massachusetts, and then ran almost parallel to the Connecticut River on a Keene-Sunapee-Whitefield-Dixville Notch line. All of Vermont lay in the heavy rain sector. Mt. Mansfield reported the greatest amount, 5.90 inches. Mays Mills in the south was a close second at 5.73 inches. Practically all sections received over three inches. Gilman in the northeast caught the least, 1.29 inches.

TROPICAL STORMS AND HURRICANES

Ex-Hurricane Belle in August 1976

Southern and western New England appeared to be threatened with a visit from a full hurricane on the evening of August 9, 1976, when Hurricane Belle stood less than 100 miles off the Delaware capes packing hurricane-force winds. But the cold surface water weakened the storm's source of energy and diminished its physical force. Winds at landfall on western Long Island during the early morning hours of August 10 did not attain hurricane force.

The center pursued a north-northeast path through New England on a Hartford-Amherst-Keene-Berlin line, rather close to Doria's advance in August 1971. Heavy rains occurred over Vermont on the 9th and 10th. Readsboro at 4.11 inches and Mt. Mansfield at four inches reported the greatest one-day totals on the 10th; very little fell before or after the main deluges that day. Amounts were lighter in the extreme northwest: St. Albans 0.72 inch and Burlington 1.23 inches.

Ex-Hurricane David in September 1979

After skirting the Florida east coast and coming ashore near Savannah, Georgia, Ex-Hurricane David followed a parabolic course that carried the center diagonally over central Vermont from southwest to northeast on September 6, 1979. The line of advance over the state ran from Danby to Plymouth to Thetford. The storm had long since lost its tropical structure over the Carolinas.

Heavy rains fell over most parts of the state. Dorset garnered by far the most with 3.80 inches on the 6th, and a storm total of 4.15 inches. Searsburg Station had 3.66 inches. Lightest amounts occurred in the southeast: Bellows Falls 0.91 inch and Vernon 1.02 inches.

Atop Mt. Washington, Ex-Hurricane David raised the greatest wind force measured there since 1942. At 12:55 P.M. on September 6, a peak gust of 174 miles per hour burst out of the southeast. The wind tipped over the body of a construction trailer, and then, after the center had passed and the wind direction reversed, set it back up again. The peak gust in 1942 had been 180 miles per hour.

Ex-Hurricane Frederic in September 1979

Hurricane Frederic was the most powerful hurricane to strike the United States coast in a number of years when it devastated the Mobile (Alabama)

In the early evening of September 14, 1979, a wedding party was rehearsing on the newly restored Bedell bridge between South Newbury, Vermont, and Haverhill, N.H., when Hurricane Frederic tore the bridge from its piers beginning on the Vermont side. With only seconds to act the group ran to the New Hampshire shore as the structure collapsed into the river. All escaped except one bridesmaid who was trapped for several hours until freed with chainsaws. The wedding took place the next day as scheduled, but it was a sad day for members of the Bedell Covered Bridge, Inc., who had worked for years to restore the 113 year old bridge. Courtesy Connecticut River Valley Covered Bridge Society.

Bay area on the night of September 12-13, 1979. Following a parabolic course similar to David's a week earlier, Frederic moved well west of that track while converging toward Vermont. The center crossed the extreme northwest portion from near Shelburne to near Troy on the Canadian border. Nowhere did rains exceed two inches. Bristol measured 1.70 inches and Mt. Mansfield 1.63 inches. Again the southeast was driest: Bellows Falls 0.52 inch and Vernon 0.55 inch.

TORNADOES

"Though our quiet State is seldom visited by tornadoes, there have been a sufficient number to show that we are not exempt."
(Dr. Hiram A. Cutting, *Climatology of Vermont*, 1877)

Vermont has never experienced a major killer tornado such as the Great New Hampshire Whirlwind on September 9, 1821, the Wallingford Disaster in Connecticut on August 9, 1878, or the Worcester County Tornado in Massachusetts on June 9, 1953. Yet many lesser tornadoes have swept over the Green Mountain area that had the potential of causing destruction of major proportions, but failed to strike any settled areas when at the height of their development.

Official tornado statistics over the years are not very reliable. In the early days, any severe windstorm might be called a tornado, a cyclone, or even a hurricane. In the post-Civil War period, it was thought that tornadoes occurred only on the Great Plains of the West, and many local storms in the East suspected of having funnels were classified as windstorms. As a result, during the early years of the twentieth century very few tornadoes were reported in Vermont. In a compilation by the U.S. Weather Bureau of tornadoes occurring between 1916 and 1958, only twelve instances were listed for the state.

The tragic consequences of the Worcester County Tornado in June 1953 and of the Waco Tornado in Texas in the previous month awakened the federal weather authorities to the tornado menace, and they instituted a system of tornado spotting and warning, along with a public education program. This has resulted in many more tornadoes being reported in the past three decades. New England weather officials, who had been prone to dismiss reports of tornadoes, began to recognize many for their true character. As a result, there have been eighteen tornadoes reported in Vermont since 1953. As Dr. Cutting wrote, Vermont is not exempt from tornadoes and warnings should be taken seriously by all citizens.

Vermont Tornado Occurrences

On June 23, 1782

By a person from the State of Vermont, we are informed that on Sunday the 23d inst[June] a terrible hurricane began in a place called Pawlet, in which place and Manchester it did great damage to the grain, buildings, &c. to the eastward of Manchester it seemed to divide into two veins, one taking a south-east, the other a north-easterly direction. The south east vein passed through Wethersfield to Claremont, and continued its course to Corydon (which was the farthest place in its course which our informant had heard) with such violence as to destroy the fruits of the earth, tear up and twist off trees, and destroy buildings of every kind that stood within its verge.

The north-east vein was a most terrible storm of hail and rain, attended with an almost incessant peal of thunder, and flashes of lightning in such quick succession, that the whole hemisphere appeared a tremendous glare of fire. The grain in its course was almost entirely destroyed; at Royalstown the rain and hail fell in such amazing quantities, that the water was knee deep in the houses, many buildings were undermined and ruined, one house was thrown down, and carried a considerable distance by the flood. Our informant says, very credible people assured him, that some of the hail stones measured six inches in length, and were supposed to weigh near a pound. The storm passed to the eastward of Royalstown, across Connecticut river, but we have received no account from those parts.

Connecticut Courant (Hartford), July 9, 1782.

On September 19, 1787, in Rutland County

The Northern post informs that on Wednesday night last, there was a heavy storm of wind and rain, in the towns of Rutland, Clarendon, Wallingsford,[sic] &c. Considerable damage was done to barns and other buildings by the wind, and great numbers of trees blown up by the roots. The freshet occasioned by the great quantity of rain that fell, was so high, as to render it impossible for the post to enter Rutland by his regular route.

Vermont Gazette (Bennington), Sept. 24, 1787.

On May 3, 1790, in Rutland

"Whirlwind at 3:30 P.M. Several buildings and trees blown down." Temperature at 2 P.M. was 78°.

Samuel Williams, Meteorological Observations, manuscript in Harvard University Archives.

The Great New Hampshire Whirlwind and Vermont Tornado Outbreak in September 1821

Tropical air covered all of New England on Sunday, September 9, 1821. Perhaps it had been brought northward by the Norfolk-Long Island Hurricane six days earlier. Noontime temperatures rose into the middle 80s and the air was very humid. By mid-afternoon thundershowers broke out in the Hudson Valley of New York and soon the line of instability and turbulence moved into the Green Mountains.

The first report of severe winds and destruction came in Rutland County, though the press account did not give the exact time of its occurrence:

> Unusual Hail Storm and Wind. — An extraordinary hail storm, attended with thunder and lightning, and in some parts with wind, amounting to a tornado, passed over Sudbury and Brandon and a part of Hubbardton and Pittsford, on Sunday the 9th inst. The hail stones fell in Brandon village and its vicinity of a remarkable size — many of them were picked up that measured from ten to fourteen inches in circumference, and nearly in globular form. It may be said that the greater quantity of the hail stones, which were sufficient to whiten the face of the earth, were from the size of a hen's egg to that of a goose egg. The hail fell with sufficient velocity to half cover themselves in the toughest green sward, and have been found to do some damage to small cattle, sheep, etc. In Brandon village the storm was attended with very little wind.
>
> In Hubbardton and Pittsford, a number of buildings were destroyed, and orchards, as well as many acres of trees of the forest, were thrown down. Many apple trees, from ten to twelve inches through, were torn up by the roots, and carried to the distance of 30 or 40 rods. — Books and other light articles, were found in the fields in Pittsford, known to belong five to six miles distant.
>
> *Rutland Herald*, Sept. 17, 1821.

The next report of tornadic activity came at Cornish, New Hampshire, on the Connecticut River opposite Windsor. Cornish and Windsor lie about 38 miles southeast of the Rutland County storm area in a straight line. Either the tornado skipped over the mountains and touched the surface again in the Windsor-Cornish area, or a new tornado funnel formed. A contemporary account of the Great New Hampshire Whirlwind appeared in the *Collections of the New Hampshire Historical Society* in 1824, and lengthy excerpts are reprinted in *Early American Tornadoes, 1586 to 1870* by David M. Ludlum.

The Great New Hampshire Whirlwind passed through the Sullivan County communities of Cornish, Croydon, Wendell, and Sunapee, cutting just north of Claremont and Newport. It crossed the northern part of Lake Sunapee, lurched over the southern spur of Mt. Kearsarge, and completed its ravages in the Merrimack Valley about 12 miles northwest of Concord. Its path extended about 23 miles through sparsely settled country. At least six persons were reported killed, and forest destruction was great. As a tornadic force in New England, it rated second only to the Worcester County Tornado in 1953, which killed ninety people and did $53 million damage.

A second path of tornadic activity ran from Washington County southeast to the Connecticut River. A "powerful whirlwind" passed through Berlin, which lies across the Winooski River from Montpelier. Later a damaging funnel struck some 28 miles to the southeast in Haverhill, New Hampshire, causing structural and extensive forest damage.

> On Saturday, 9th instant., a powerful whirlwind passed from west to east, the distance of about two miles. In Berlin, twisting potatoe tops, bushes and brakes, close the ground, and carrying them up entirely out of sight. In its progress it crossed Stephen's Branch and entered Onion River, raising the water in a body, about the circumference of a barrel, and carrying it in a column to the clouds. The interesting phenomenon was witnessed by Mr. Jacob Davis, Jr., and a number of others, by whom we are favored with the fact.
>
> *Vermont Watchman* (Montpelier), Sept. 25, 1821.

Tornado at West Fairlee on August 31, 1825:

> I am told by inhabitants who have lived in town longer than the writer, that a most marvelous freak of nature occurred on the night of the 31st of August, 1825, which may be worthy of notice here.
>
> A disastrous tornado or (as the inhabitants call it), whirlwind, swept over this section the night above stated. It commenced in Strafford, running easterly to Connecticut River. Its mean width was about 125 rods. It passed through here about eleven o'clock. The night was dark beyond description. It was accompanied with terrific thunder and hideous lightning, unroofing buildings, and in some instances shattering them in a thousand pieces, and leveling forests to the ground wherever it traveled.
>
> Abby Maria Hemenway, *The Vermont Historical Magazine*, 2, 914.

On August 9, 1829, at Barnet and Peacham

A tornado of undetermined size swept the forest areas of northern New York State, churned across Lake Champlain, bounded over the crest of the Green Mountains, and descended to the earth's surface very close to the Connecticut River in eastern Vermont. Extensive damage was reported by the *Essex County Republican* in the afternoon of 9 August on the New York side of Lake Champlain. Houses were blown down, trees of 18-inch diameter snapped off, and shipping on the lake suffered downed masts and torn sails. A hay stack of considerable size was lifted up and carried eastward about a mile and a half out into the wide lake.

The whirl apparently exhibited a skipping action, or it dissipated temporarily, as the press of Vermont failed to report its presence or any other tornadic action until villages close to the Connecticut River were reached by the turbulent front. A southeastward track from Plattsburgh to the villages of Barnet and Peacham in Caledonia County would have carried north of the population centers of Burlington and Montpelier. The St. Johnsbury *Farmers Herald* told of the "tremendous whirlwind" in its vicinity:

> This whirlwind commenced in the south west part of Peacham at 5 o'clock p.m. of the 9th inst. Immediately after a heavy shower of rain with thunder, the clouds were observed to be in violent agitation. As it advanced it took down an old barn in Peacham, belonging to Mr. O. P. Chandler. Pursuing an easterly course, it twisted off trees 2 ft. in diameter. Its breadth was from 10 to 35 rods. It took down a barn on the bank of Harvey's pond in Barnet. The air appeared to those who were near to be full of timber; two sheep and a colt were carried into the pond, together with shingles, hay, etc. and a column of water was raised from 100 to 200 feet into the air. It destroyed Mr. Brock's orchard, and unroofed Mr. W. Gilfillan's barn and sheds; and took off the roof of Mr. Lang's house and carried off two beds and bedding and other furniture of the chamber, which has not yet been found. It also prostrated a barn at the same place. A man was carried 6 or 8 rods and struck several times in his course by falling timber, and at last was stopped by the fence much bruised. It is singular that the timber &c. of the house were carried northwesterly tho' the direction of the whirlwind was the reverse. It is also a fact, that where it passed through forests, more timber fell north and northwest than in an easterly direction.
>
> *The Farmers Herald* (St. Johnsbury, Vt.), Aug. 19, 1829.

On July 2, 1833, at Holland:

> On the 2d of July, 1833, Holland was visited by a violent tornado. It commenced on Salem Pond in Salem, and passed over the town in a northeasterly direction. It was from half to three quarters of a mile wide, and it prostrated and scattered nearly all the trees, fences, and buildings, in its course. It crossed the outlet of Norton Pond, and passed into Canada, and its course could be traced through the forests nearly to Connecticut River.
>
> John Hayward, *A Gazetteer of Vermont* (1849).

On July 3, 1842, at Victory:

> Upon the 3d of July, 1842, there was a tornado in the town of

Victory (Essex County) of remarkable force. Its devastations commenced upon the top of a high hill in that town, where its path was only a few yards wide, but it gradually increased to about a half mile in width, sweeping everything before it for about two miles, when it seemed to lose its power. Its track was a forest, yet it not only tore up the trees, but the soil also, piling with twisted and broken trees in huge rows near the place where its fury seemed spent. The noise of this tornado was heard for more than ten miles, and was supposed by many to be an earthquake. It was accompanied by heavy thunder and incessant lightning, with torrents of rain.

Hiram A. Cutting, *Climatology of Vermont*, 12.

On September 20, 1845, at Burlington

This town was visited this evening by the most grand and awful storm of thunder, lightning, rain and wind that we have ever experienced.

The wind during the day was south, but toward night it veered around to the east, and so on to the North and Northwest. The clouds gathered blackness, and on the lake to the NW., the appearance was like the premonitory symptoms of a storm at sea. At dusk, the horizon in the West from South to North, and so around to East, was alternately, with the rapidity of thought, a blaze of fire or a sheet of blackness, dark as midnight.

In the distance could be heard the roaring as of mighty waters.

Cyclones are not frequent in Vermont, but they occasionally strike as they did in Fairfax on August 7, 1907. This pile of rubble was all that was left of the Ryan's large barn after the cyclone departed. Forty-five cattle died when the barn collapsed.

153

Onward, fitful and furious, came the raging blast; trees are levelled to the ground, chimneys demolished; barns and houses unroofed; the fragments flying all around show the power and fury of the storm. The rain descends in very torrents. The balustrade and the chimneys on Messers. Strong's store (formerly the Burlington Hotel) and the chimneys on Messers. Peck's store, are swept off clean to the roof.

The roof of Mr. John Bradley's new brick barn was entirely blown off and the building nearly demolished. Mrs. Doctor Moody's fine dwelling, near the female seminary, had the roof entirely taken off, and the storm of wind and rain poured in upon the terrified inmates; the house and everything in it were completely drenched.

This is something new for Burlington. I have not as yet heard of farther damages.

The steamer Burlington left the wharf a short time before the blow – in time, I judge, to get into Shelburne Bay, or over to the other shore, before the storm.

Correspondent of the *Commercial Advertiser* (New York), from Burlington, Sept. 20, 1845.

This was the tail end of the great Adirondack Tornado of September 1845 which originated near the shore of Lake Ontario west of Rochester and pursued a course across that lake, through the entire breadth of the Adirondack Mountains, and over Lake Champlain from south of Port Kent to Burlington. A windfall of downed trees could be traced across much of northern New York for many years.

On October 2, 1859, at Burlington

A Water Spout in the Bay. The interesting and uncommon phenomenon on this lake, of a *water spout*, was witnessed near the breakwater in the Bay during the squall about 4 o'clock on Sunday afternoon last [Oct. 2]. Our townsman, Henry R. Campbell, Esq., who saw it, with a number of other persons, describes it as an elevation of the water in a cylindrical column with a convoluted or twisted surface and irregular top, apparently about ten or twelve feet in diameter. It passed slowly toward the North, striking the beach toward Sharpies Point, and throwing about the sand and gravel at a great rate.

The Daily Free Press (Burlington), Oct. 5, 1859.

On May 13, 1866, at Barnet

A great tornado about six and one half o'clock P.M. The wind was in the south and for about one hour there was every appearance of a heavy thunder shower. The clouds came rolling up over the hills & were black as night. There was a little but not severe thunder and lightning. Soon the wind began to blow with great severity, taking large trees up by the roots, twisting off the tops of others, unroofing some barns, blowing down others, as well as some houses. It totally demolished the toll bridge across the Connecticut River at this place. It seems to have extended a little more than a mile in width. It was more severe however in some places of its course than others, & some places it seemed to widen. After the tornado passed there was quite a heavy shower of rain & a very little hail.

Smithsonian observer, Barnet, in National Archives,
Washington, D.C..

On September 30, 1907, at Enosburg Falls

On September 30, about 3 P.M., a severe tornado, or local storm, passed within 5 miles of this place. It was accompanied by a peculiar funnel-shaped cloud and a roaring sound, and left a well-defined track of about 5 or 6 rods in width for a mile or more. At one set of farm buildings in the path of the storm, the chimneys were blown down, the shingles blown from one side of a large barn, and the outbuildings demolished.

L. H. Pomeroy, Climatological Report: New England Section, 19-9,
September 1907, p. 68.

On February 12, 1954, at Burlington

Scattered clouds at 800 feet, scattered clouds at 1500 feet, barometer 1025.3 mb., temperature −8°F, dew point −21°F, wind north 18 mph, gusts to 27 mph. Series of funnel-shaped clouds over Lake Champlain mostly extending from clouds to surface and moving from north to south. Funnels appear to be small tornadoes forming then moving southward a few hundred yards, dissipating, then new ones forming at north end of cloud cover just west of station.

Robert G. Beebe and J.E. Stork. Winter Funnel Clouds
over Lake Champlain, Weatherwise, 8-2 (April 1955), 44-46.

On May 20, 1962, at Bakersfield-Westfield-Albany:

A series of at least three tornadoes was spawned in an eastward moving area of severe thunderstorms which moved across northern Vermont. Thunderstorm wind squalls caused property damage from St. Albans eastward. The first of the three known tornadoes, each of which moved to the northeast, touched down at Bakersfield to cut a wavering path of 40 to 60 yards wide and 3 to 4 miles long, at about noon, EST. The next hit Westfield about one-half hour later. The third twister apparently made two contacts with the surface. The first was at Albany, at about 12:45 P.M., and the second at Coventry, 12 miles to the northeast. Destroyed were farm buildings, a trailer home and trees. Damaged were additional homes and buildings. Utility wires serving some communities were broken. Local heavy rains accompanying the thunderstorms caused washing and flooding in some areas, affecting tilled land, roads, and cellars.

Storm Data (May 1962): 52.

On July 9, 1962, in Windsor County

At least 2 tornadoes were spawned by the same eastward moving cell in this part of the state. The first one touched down at a number of wooded areas along its 16 mile course from Chester to Weathersfield. Storm moved from the southwest to the northeast with average forward speed of about 25 miles per hour; time of first touchdown was 9 A.M. The second tornado made contact with the surface at Springfield shortly after 9 A.M. Its path was also northeastward and about 5 miles long to the Connecticut River where it crossed and continued into New Hampshire for an additional three miles. This tornado made frequent contact with the surface in its path and caused extensive damage to buildings in Springfield, Goulds Mills, and in several other villages. Trees were twisted off or uprooted and several communities lost power and phone services.

Storm Data (July 1962): 89.

On July 9, 1962, at Addison County

Funnel cloud observed by pilot at 12:30 P.M. about 6 miles northeast of Vergennes. Height of funnel estimated at 1,000 feet or more above ground.

Storm Data (July 1962): 89.

On August 3, 1970, at St. Albans

First seen as waterspout on Lake Champlain at 3:25 P.M. Two boats damaged at Maquam Bay. Moved on land and hit a camp home in St. Albans Town, injuring 7 occupants. Funnel lifted and then dipped again farther inland to demolish a set of farm buildings, including a new barn. A heifer was killed. Path in general east-northeast direction. Funnel lifted several times. A second funnel was sighted in the area but did not touch ground. Radar indicated a possible funnel later 30 miles to the east, but no further reports of damage.

Storm Data (August 1970): 131-32.

156

SNOWSTORMS

A snowfall consists of myriads and myriads of minute ice crystals whose birthplace lies in the subfreezing strata of the middle and upper atmosphere. At the core of every ice crystal is a minuscule nucleus, a solid particle of matter of varying composition, around which moisture condenses and freezes. Liquid water droplets floating in the supercooled atmosphere and free ice crystals cannot coexist within the same cloud because the vapor pressure of ice is less than that of water. This enables the ice crystal to rob the liquid droplets of their moisture and grow continuously by accretion. The process can be very rapid, quickly creating ice crystals, some of which adhere to each other to create an aggregate of ice crystals — a snowflake.

Simple flakes possess a variety of beautiful forms, usually hexagonal, although the symmetrical shapes reproduced in most photomicrographs are exceptional specimens and not usually found in most snowfalls. Snowflakes found in a typical snowfall consist of broken fragments and clusters of adhering ice crystals and only infrequently are the original six-sided figures retained.

For a snowfall to continue once started, there must be a constant inflow of moisture to feed the hungry condensation nuclei. This is supplied by the passage of an airstream over a water surface where it picks up moisture and carries it to a higher region of the atmosphere. For Vermont, the Atlantic Ocean is the chief source of water vapor, with the Gulf of Mexico a secondary contributor for storms traveling a long distance. Some moisture is carried eastward from the Great Lakes, but once these lakes freeze this minor source of moisture is cut off.

A snowstorm is a meteorological happenstance. Many elements in the

157

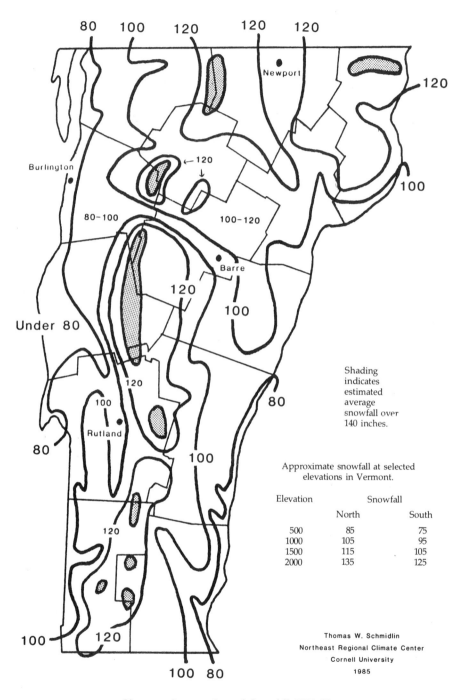

Vermont Average Annual Snowfall 1951-80

Shading indicates estimated average snowfall over 140 inches.

Approximate snowfall at selected elevations in Vermont.

Elevation	Snowfall	
	North	South
500	85	75
1000	105	95
1500	115	105
2000	135	125

Thomas W. Schmidlin
Northeast Regional Climate Center
Cornell University
1985

THE VERMONT WEATHER BOOK

atmosphere must be brought together in the right proportion at the right time for one to occur. The main ingredients are moisture and cold. There must be a continuous flow of a moist airstream over a substantial period, along with steady cold of below-freezing degree. The cold is usually borne to the scene with a northeasterly or easterly air flow at the surface, and at upper levels the moisture comes from the southeast or south. If the surface flow should come from the southeast, or shift there, the snow frequently turns to rain and creates a slushy condition on the ground.

The thermodynamics and hygrodynamics of large cyclonic storms govern snowmaking through the successive processes of convection, cooling, condensation, and finally precipitation. In addition to general snowstorms covering a wide area, the lee shores of substantial bodies of water often initiate snowfalls when cool airstreams flow over these warm bodies of water. This produces the lake-effect snow squalls that make the Great Lakes a famous snow belt, and the same process can produce a lake-effect snowfall along the shores of Lake Champlain and the rising terrain back from the lake shore. The mountains, too, can provide the needed lift to create snow squalls when moist air passes over them (this is termed orographic precipitation). Snows produced by orographic precipitation may amount to only a few flakes, or a single flurry, but on occasion may mount to several inches, much to the consternation of the forecaster and the surprise of the public.

More so than a snowstorm, winter's worst menace to Vermonters is a freezing rain that coats all outside objects with an icy sheath known to meteorologists as glaze. Freezing rain occurs when water droplets fall from an above-freezing layer of air aloft through a shallow layer of below-freezing air at the surface of the earth. Upon impact or shortly thereafter, the droplets freeze on all exposed objects, coating them with a varying thickness of glaze ice. If the layer of cold air at the surface is deep, the rain droplets freeze during their descent and form small ice pellets, sometimes called sleet. Occasionally, the two forms fall at different times during a storm, or may be intermingled, congealing all into an icy mass covering the ground and clinging to wires and antennas. Despite the damage and inconvenience caused by an ice storm, it presents a most spectacular scene when the rays of the rising sun glint on ice-coated trees and shrubs with thousands of gleaming spangles.

April Fool's Day Snowstorm in 1807

Snow! Snow! Snow! On Tuesday morning the 31st of March we were visited by a snow storm which continued with little intermission till Friday morning, four days: during this time the snow fell about two and a half feet—which, with the snow on the ground before the storm, made about six feet upon a level, in the woods, and more than four feet in open land. The fences in many places were entirely covered, the tops of the posts could not be seen. On Saturday the fourth inst. it was a gloomy sight indeed to behold such a body of snow on the ground. We have since, however, been visited with the sun, for a few days, which has lessened the snow considerably. For a fortnight we did not receive any mail from the southward, in consequence of the great snow; there has been very little passing.

We think our readers will ask no other apology from us for getting out no paper last week.

With these words the editor of the *Weekly Wanderer* at Randolph took notice of one of the most memorable snowstorms in the early history of the state. A storm of great physical dimensions, it extended over all of New England with very heavy precipitation for the season and severe winds. The center of the cyclonic disturbance passed near New York City about noon on March 31, 1807, and then cut across southern New England on an east-northeast track. At New Haven the barometer sank to a reading of 28.75 inches, indicative of a storm of unusual pressure depth. Located to the north of the center's path, Vermont was in a zone of heaviest snowfall that usually lies from 100 to 150 miles to the northwest. Accounts from the Vermont press were unanimous in acclaiming the severity of the wind and the depth of the snow.

The editor of the Danville *North Star* on April 7th took measure of the snow situation:

The unusually severe storm of last week, blocked up the roads to such a degree, that they are in many places in this vicinity impassable with horses or sleighs. . . . We presume there never was such a time known here at this season of the year. The snow is considered to be five feet deep on a level, in many places covering the fences, and rendering it extremely difficult to travel unless it is with snow shoes.

Over at the newly established capital of Montpelier, the editor of the *Vermont Precursor* commented in like vein:

On Tuesday last (March 31st), there was a snow storm, the most driving we have had this winter, though not cold, the mercury in the Thermometer being at the freezing point. It snowed most of Friday and Saturday (April 3d & 4th). The snow is now near four feet deep in the streets and common roads on a level. So much snow on the ground at this time of year in this place, it is said, has never been known before, since the country has been settled.

When writing his "Natural History of Essex County" in the 1870s, Dr. Hiram Cutting of Lunenburg in northeastern Vermont singled out this storm for special mention:

There was on the last day of March a great snow storm accompanied with a very high wind – such a wind as is seldom known. It block-

THE VERMONT WEATHER BOOK

aded the roads so that they were not passable for some days. On the first day of May [April ?] the snow would average 4⅓ feet deep, and the weather cold and forbidding—yet warm days soon came and crops came forth with great rapidity, and it is seldom that a better harvest is gathered than in that year.

Abby Maria Hemenway, *Vermont Gazetteer.*

The Great Late March Storm of 1847

After the Great March Blizzard of '88, a Rutland editor recalled experiencing a similar storm as a young man at about the same time of year. He searched the files of the *Rutland Herald* and uncovered an account of a super storm occurring on March 26-27, 1847, with a time-honored reference to a respected local authority: "The oldest inhabitant has just dug out, and he assured us that he has not seen the like before, and with this positive assurance on his part we will close this notice."

The 1847 account related that the storm commenced on Friday evening, the 26th, with rain falling until about 10:00 P.M. when the wind shifted to the north and snow took over. It continued snowing all night and through the next day until evening. No depth was given for Rutland, but highway communication was blocked for several days.

Professor Zadock Thompson measured the storm's behavior at Burlington with his array of meteorological instruments. His barometer commenced falling rapidly about noon on the 26th with some snowflakes already in the air and the wind coming out of the northeast. Bluebirds and robins had appeared on the 25th and snow buntings on the 26th, so few were prepared for the wintry scene to follow. Thompson's thermometer fell from a high point of 39° at noon on the 26th to 32° at 9:00 P.M., leading one to believe that much of the early snowfall must have melted. A north wind prevailed during the period of heaviest snowfall when his barometer dropped almost an inch in twenty-four hours. In measuring 17 inches of new snow, Thompson commented: "a driving storm. Roads all blocked up...The snow on the 27th was very heavy and yet very badly drifted. Mail delayed by it two days."

Dr. Hiram Cutting at Lunenburg included this in his list of big snowstorms, but gave the date as 1849 (believed to be a typographic error): "Twenty two inches fell and the roads were blocked up with drifts." At St. Johnsbury, the snow fell 12 inches deep in the valley and 18 inches on the highland, covering all the fences, according to *The Caledonian.*

At Randolph, William Nutting noticed a rain with a southeast wind on Friday evening, turning to snow before midnight: "Snow storm continued all night and today from N.E. & N. with strong wind & drifted

161

all into heaps. By estimation 15 inches." The next day was "fair but very blustery—no meeting—roads impassable."

The historian of Barnard, W. M. Newton, described the storm of March 27, 1847, as "the greatest snowstorm in the history of the town, totaling 30 inches by actual measurement."

The Snowy April of 1874

"A cold, stormy, backward spring. April 1874 will be noted for its large quantity of snow and severity of storms," wrote the Wilmington correspondent of the *Country Gentleman*, C. T. Alvord, who resided on the high ground along the Brattleboro-Bennington highway. He measured a total of 54 inches during the first spring month, compared with an average during the past years of only nine inches. On May 2, Alvord found 18 inches of snow in the woods, and on the 21st some still remained when he made a trip to nearby Searsburg.

At Woodstock, snow fell on thirteen days for a month's total of 49 inches. The greatest storms were concentrated near the end of the month: 15 inches on April 25-26 and 10 inches on 28-29. Snow lay on the ground at Woodstock to a depth of 31 inches on May 2. Additional light falls came on May 8, and the ground was found to be frozen to a depth of one foot on May 7. The total snowfall for the entire season was a very substantial 156.2 inches.

At Strafford in Orange County, 12 inches fell on both the 26th and 29th, and the last traces of snow did not disappear until May 27.

The St. Johnsbury observer reported "nearly three feet" on the ground on May Day, and at Craftsbury "more than two feet" discouraged any celebrants of the spring festival.

The First Blizzard of 1888: January 25-26

"The present storm is pronounced the severest ever known in this section of Vermont, the wind blowing a gale and the highways completely blocked," wrote the editor of the St. Johnsbury *Caledonian*. The observer at Strafford in Orange County agreed: "The blizzard of the 27th was the hardest in memory of the oldest inhabitant; the roads remained blocked in many places so that business was suspended for three days."

Whether or not the railroads can function is an indicator of the severity of a storm. The Burlington *Free Press* reported that huge drifts in cuts at Shelburne and Charlotte shut down the Rutland Railroad from Thursday night on the 26th until Sunday morning the 29th. Ten locomotives

More than a century has passed since the great blizzard of '88, but Vermont weather watchers still consider it the wildest snowstorm in the state's history. In this post-blizzard Brattleboro scene, the mailman makes his appointed rounds despite the storm, in the best tradition of the U.S. Post Office.

were snowed in on this stretch of track when they tried unsuccessfully to snowbuck a stalled train out of the huge drifts and became stuck themselves.

No New York mail was received at Burlington from Wednesday evening until late on Saturday when the rail block was circumvented by engaging a horse and sleigh to move the sacks. The little Woodstock Railroad took seven hours to make the trip uphill from White River Junction: "never before has there been any detention of trains of consequence from snow," declared the editor of the local *Vermont Standard*. From all corners of the state came the opinion that this was the most serious blockade in the forty-year history of railroading in the Green Mountain region.

The blizzard conditions resulted from an unusual storm track that unfolded on January 25 and 26. An energetic low-pressure area looped southeast from the Great Lakes, crossed central New York State and Massachusetts, and then turned northeast to the Gulf of Maine where the storm developed a very low barometric pressure of 28.60 inches at Eastport, Maine. Gale and whole-gale winds blew the length of Vermont during the 26th, sweeping the snowfall of 10 to 22 inches into huge drifts, and then continued to rage all the next day, defying attempts to open highways and clear railroads.

The snowfall began at Burlington at 4:00 P.M. on the afternoon of the 25th. A southeast wind prevailing earlier in the day raised the thermometer from a morning $-7°$ to an evening $11°$ soon after the start of

163

the snowfall. A snow period of 27.5 hours dropped 12 inches of new snow on the Queen City. Temperatures on the 26th ranged from a morning 18° to an afternoon 12°.

In the southwest corner of the state, Manchester reported the first flakes at 7:00 P.M. on the 25th, with the mercury at 14°, up from a morning low of −12°. The snow continued to fall for twenty-two hours.

In the central section, the mercury at Northfield contracted to −30° at sunrise on the 25th, and climbed to 3° when the snowfall began at 9:45 P.M.

The 26th developed into a true blizzard day in all sections of the state. Arctic airstreams, drawn south by the intense circulation around the deep low-pressure center over Massachusetts, picked up the fine flakes from the ground and drove them along parallel with the surface. The wind at Brattleboro mounted to 47 miles per hour from the northwest at midday. The barometer dropped to 29.46 inches when the storm center passed a short distance to the south while still deepening.

The greatest snowfalls were in the northeast where Newport had a new fall of 22 inches on top of the 15 inches of the week before. This was a final contribution to the record 45 inches that fell during January.

The storm's impact at Northfield was typical of the interior valleys. The new fall of 22 inches produced a record depth of 54 inches at the end of the month. The severe conditions at this locality may be appreciated from the temperature readings of 6° at 7:00 A.M., 7° at 3:00 P.M., and 5° at 10:00 P.M. on the 26th. During this period the wind mounted to a peak of 42 miles per hour from the north, producing a severe windchill. Other storm amounts were: Montpelier 19 inches, Waterbury 18, Bradford 15, Woodstock 17, Burlington 12, and Lunenburg 10.

The Great March Blizzard of 1888

No paths, no streets, no sidewalks, no light, no roads, no guests, no calls, no teams, no hacks, no trains, no moon, no meat, no milk, no paper, no mails, no news, no thing—but snow.

Bellows Falls Times, Thursday forenoon, March 15, 1888. (A parody of "No!" by Thomas Hood [1799-1845])

The Blizzard of '88 stands unique in the storm annals of the northeastern corner of the United States. No other atmospheric disturbance has brought together the combination of deep snowfall, gale-force winds, and bitter cold, the three ingredients of an eastern blizzard, in such an extreme degree over such a wide area as did this late-winter snowstorm

164

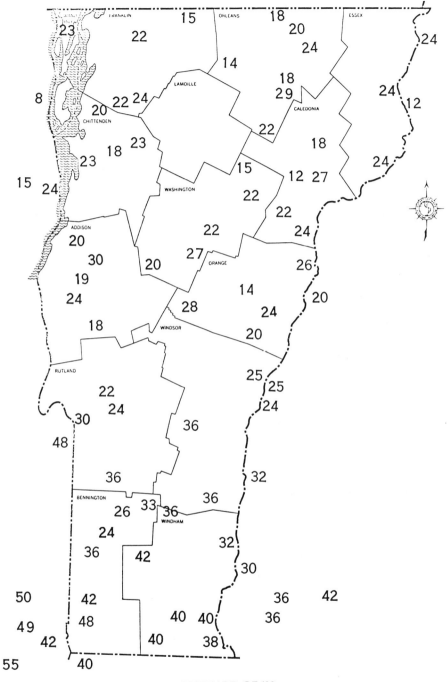

BLIZZARD OF '88
Snowfall: March 11-14, 1888

SNOWSTORMS

coming within ten days of the spring equinox. Raging east of the Appalachians from North Carolina to Canada and out over the ocean, the Blizzard of '88 claimed hundreds of victims on land and sea.

The Weather Map

The appearance of the weather map over the central and eastern United States on the morning of Saturday, March 10, presented nothing to intimate that the greatest snowstorm of the century was in the making. A trough of low pressure of modest dimensions and strength over the Mississippi Valley was sandwiched between two high-pressure areas whose centers lay in Canada: one to the northwest over the frozen tundra west of Hudson Bay, the other over the Gulf of St. Lawrence and Newfoundland. The interaction between the three atmospheric features was to produce blizzard conditions and bury the Green Mountains under the greatest total mass of snow ever recorded in one storm.

On Sunday morning, the 11th, when the eastward-moving trough reached the Atlantic Coast, a cyclonic circulation formed off the Carolinas, and twenty-four hours later it was centered as a severe storm off the Delaware capes. To the east of the storm center, a vast warm sector developed with tropical air streaming northward, bringing warmth and moisture. Meantime, the anticyclone over central Canada developed great magnitude and advanced to the Great Lakes region while sending record cold airstreams eastward as its contribution to the developing storm system on the Atlantic Coast.

Inland about 15 to 20 miles, however, the moist air flow from the ocean encountered a resisting wedge of very cold, continental air directed eastward by the high-pressure area then over the Great Lakes. The maritime air was forced to glide up and over the more dense air, hugging the surface as far east as central Massachusetts and New Hampshire. When forced to rise, the oceanic current cooled and condensed its enormous load of moisture in the form of snowflakes. The steadily rising terrain west of the Connecticut River augmented the upslide motion and wrung additional moisture from the clouds. With the low-pressure center stalled along a line of latitude just south of Cape Cod, the moist airstream continued to flow, extending the precipitation period to about forty-eight hours, or close to double that of a normal winter coastal storm. This accounts for the unprecedented amounts of snow reported across a zone of southern New Hampshire and Vermont and in adjacent portions of Massachusetts and New York.

On Tuesday morning, the 13th, all of Vermont except the northern border communities lay in the cold wedge of air with temperatures below

10°. The lowest reading at a regular weather station reached 4° at Manchester in the southwest, while Brattleboro in the southeast reported 6°. At the same time, Newport, north of the cold wedge of air, dropped only to 21° and rose above freezing in the early afternoon when maritime air arrived from the east. The leading edge of the cold air did not penetrate beyond a north-south line running through central New Hampshire and Massachusetts. Readings on the Maine coast remained in the 30s, and Boston's thermometer hovered around 32°, causing mixed precipitation for two days.

Light snow began over parts of Vermont about noon on Sunday, the 11th, but did not develop unusual intensity for many hours. Meantime, the high-pressure area over the Atlantic provinces of Canada made itself felt by blocking the normal northeast movement of the storm center, causing it to stall south of Long Island early on Monday, the 12th. Caught as it was between the two high-pressure systems, the snowstorm whirled around like a trapped dervish, all the while lowering its barometric depth and increasing its wind force. Snowfall increased over Vermont from moderate to heavy as more and more Atlantic moisture was carried inland over the wedge of cold air that was growing colder and colder.

During the daylight hours of Tuesday, the 13th, the storm center performed a tight, counter-clockwise loop over Rhode Island and adjacent waters. During the many hours required for this operation the precipitation continued over inland New England, giving the region about double the amount of precipitation as would fall in a typical heavy snowstorm.

Most Vermont stations reported excessive amounts of precipitation for a winter storm. When melted to obtain its water content, Vernon measured 4.35 inches, Jacksonville 4.13 inches, and Marlboro 3.61 inches—all stations in the southern zone of heavy snowfall. Most observers complained of the difficulty in getting a representative measurement on account of the heavy drifting attending the storm.

Soon after the Blizzard of '88, the New England Meteorological Society sent a circular to all postmasters in the storm area asking for a report of local snow depth. An article containing this data and a description of the meteorological conditions of the storm appeared in the May 1888 issue of the *American Meteorological Journal*. The greatest depth of 48 inches was reported by Bennington. Most stations along the southern border were in the vicinity of 40 inches, and totals increased westward to a maximum of 55 inches at Troy, New York. The least amount in Vermont was 12 inches at Danville in the north. Variations for stations within a few miles of each other occur in many storms, especially when high winds are a factor. Usually a location on a mountain slope facing to the northeast or east will receive the greatest amounts in this type of storm as a result of local uplift of the moist airflow.

Around the State

Woodstock: "The most remarkable snowstorm that passed over Woodstock and vicinity for sixty years" was the lead headline in the *Vermont Standard*. The snow began about noon Sunday and continued until noon Tuesday. The wind remained at south and southeast until early Monday, when it changed to northeast and increased to a gale by evening. At 9:00 A.M. on Monday there were 9 inches of new snow and at 9:00 A.M., Wednesday, the total fall measured 32 inches. At that time, 78 inches of snow lay on the ground, or 6 feet 6 inches. No mail arrived from the south until after Thursday.

Newport: Snow commenced to fall at 1:00 P.M. on Sunday, the 11th, and continued until 1:00 P.M. on Wednesday. A total of 18 inches was reported. Temperatures were above freezing on the 11th, fell to a low of 19° at 2:00 P.M. Monday, and rose to 35° twenty-four hours later. Wind was south on the 11th, north on the 12th and 13th—the highest wind was 15 to 29 miles per hour.

Burlington: Sunday, the 11th, was a mild day at Burlington with a maximum temperature of 38° and a minimum of 33°. Snow began at 8:00 P.M. and continued for almost 48 hours, ending at 6:30 P.M. on Tuesday. A north wind on the 12th drove the thermometer down to 9° during the evening and to 5° early on the morning of the 13th. A difference of 16 degrees existed across the mountains from Burlington to Newport, a distance of 60 miles. Twelve inches of snow fell at Burlington on Monday, and 11 inches on Tuesday, for a total of 23 inches. Thermometers rose above freezing on both Wednesday and Thursday.

Manchester: Snow commenced at 10 A.M. on Sunday, the 11th, with temperature just above freezing. The wind shifted to the northeast, north, and northwest on Monday, blowing at 30 to 39 miles per hour. The thermometer read 5° at the Monday and Tuesday morning observations. A heavy snowfall continued until 8:00 P.M. on the 13th by which time 24 inches had accumulated. Much greater amounts were reported a short distance to the south.

Bellows Falls: No teams passed through the square on Tuesday. Boston papers of Monday, along with New York papers of Saturday, arrived Friday afternoon. Heavy drifting attended the storm: "Fronting the west side of Towns' Hotel, 100 feet of this drift was 22 feet wide and averaging 10 feet in depth, and 12 feet deep at its highest points...In some instances drifts were 10 to 15 feet deep... Greatest snow blockade ever experienced in this section." (*Bellows Falls Times*).

Brattleboro: Walter H. Childs of the Estey Organ Company had recently acquired a set of meteorological instruments and was able to record the conditions attending the severe storm. The barometer dropped from a high of 30.58 inches on the 10th to a low of 29.52 inches on the 13th.

Whether this was the absolute minimum is not known. His anemometer registered 62 miles per hour (corrected to modern standards) from the northeast on the 12th and 40 miles per hour (corrected to modern standards) on the 13th, also from the northeast. The thermometer slumped from a high of 40° on the 11th to a low of 6° on the morning of the 13th. Snow fell from Sunday afternoon to Tuesday morning: three inches on Sunday, 23 inches during the day Monday, 10 inches Monday night, and a final 4.5 inches on Tuesday—a total of 40.5 inches.

End-of-the-Century Storm in March 1900

A zone of very heavy snowfall spread from the Midwest, through the lower Great Lakes, and across northern New England from February 28 to March 2, 1900. Local records for the greatest amounts in a single storm and in a twenty-four-hour period were set at Lawrence, Kansas; Toledo, Ohio; Rochester, New York; and Northfield, Vermont. At the latter a total of 31 inches fell in forty-eight hours, the deepest at one time ever measured at the weather station there during its existence from 1887 to 1943. Depths of two feet were reported by the press in Orleans County, and the official measurement by the government observer at Newport totaled 22 inches.

A total of 21 inches descended on Burlington, beginning at 5:30 A.M. on the 1st and ending at 4:00 P.M. on the 2nd. An interval of light rain intervened on the late evening of the 1st. Temperatures remained well below freezing on the first day, ranging from 17° to 27°, and held just below freezing on the second, from 24° to 32°. Northfield received 26.5 inches on the 1st and 3.5 inches on the 2nd, for a total of 31 inches, its greatest single snowfall.

The center of the wide-ranging storm remained well south of Vermont. From near Nashville, Tennessee, on the evening of February 28, it advanced to the vicinity of Washington, D. C. On the next morning, twenty-four hours later, the center was located near Portsmouth, New Hampshire, where the barometer read 29.10 inches. Northern Vermont lay in the zone of heaviest precipitation and temperatures were low enough most of the time for snow making.

Big January Storm in Northwestern Vermont in 1934

"More than two feet of snow blankets city. Heaviest fall in the history of the United States Weather Bureau gives Burlington fairy-land appearance on Sunday," announced the headlines of the *Free Press* on January 15, 1934.

Not only was the total of 24.8 inches the Queen City's greatest single snowstorm to that date, but the intensity rate of 24.2 inches within twenty-four hours still stands as a record.

The storm track was typical of storms giving a deep snowfall at inland points of New England. From its origin near the Texas coast on the morning of January 11, the center crossed the Gulf of Mexico in the next twenty-four hours to a position near Mobile, Alabama, and in another twenty-four hours it reached the coast of Maryland. During the next twelve hours, daylight of the 13th, the forward pace of the storm system slowed somewhat as the center cut across Long Island and made a land traverse across southeastern Massachusetts. This put the northwestern portion of Vermont in the zone of heaviest precipitation.

Burlington's Most Intense Snowstorm

Burlington and nearby towns spent Sunday digging out from beneath a record-breaking snowfall that blanketed this section with more than two feet of snow Saturday night and yesterday morning.

The snowstorm brought plenty of trouble, but no major damage, to communication, transportation, and electricity lines, as well as putting hundreds to work on the Sabbath. Snow removal will cost the city $4,000 it was estimated.

Practically all the snow fell within a twenty-four hour period, burying this vicinity beneath the heaviest mantle in the history of the United States weather bureau station here. Milton W. Dow, meteorologist, reported last night that the total fall amounted to 24.8 inches.

The nearest approach to this record, reliable data dating back to 1883 on file at the weather bureau reveal, was on March 28 in 1919 when there was a snowfall of sixteen inches. Little snow was left on the ground by last Saturday after an early mild January had brought above-freezing temperatures nearly every day. . . .

Residents of Burlington and vicinity went to bed Saturday night while huge flakes, moisture laden, fluttered downward, covering ground, trees, bushes, streets, sidewalks, and buildings alike with a white mantle of ever-increasing thickness. They awoke Sunday morning to find their world displaying the most wintry scene within their memory.

Burlington Free Press, Jan. 15, 1934.

Vermont's Deepest Snowstorm in Modern Times: March 2-5, 1947

Stations in southern Vermont generally receive less snowfall during a winter season than do locations in northern Vermont since the latter averages about four to five degrees colder. On occasions, however, high elevations in the south often receive more total precipitation; if the temperature is below freezing, much of the precipitation may fall as snow.

A unique weather situation developed during the first days of March 1947, when a very moist airstream from the Atlantic Ocean crossed the higher elevations of the south while temperatures hovered just below the freezing mark. Since the pressure system reponsible for the onshore circulation moved very slowly over a three-day period, the oceanic airflow continued long enough to produce Vermont's deepest snowfall from a single storm in the present century.

The weather map on March 1, 1947, placed a cold front across the Gulf of Mexico on a northeast-southwest orientation, with several small low-pressure disturbances forming waves along the front. The whole system moved northeast on March 2, one small low-pressure center heading for the lower Great Lakes, another to the east, skirting the Atlantic coast line toward Long Island. These low-pressure centers merged into a single cyclonic circulation over southern New England on the evening of March 3, with rapid deepening of the center taking place. Barometers in the central Connecticut River Valley soon plunged to near-record depths for a non-hurricane situation: Hartford 28.65 inches, Amherst 28.63 inches, and Hanover 28.64 inches.

Instead of continuing on the usual northeast track toward the Gulf of Maine, the vigorous storm center curved cyclonically westward, passing over Vermont, across extreme northeastern New York State, and into the Province of Ontario. A slow-moving cold front extended southeast into the ocean off Cape Cod and the Islands. Lying to the north of the cold front, southern Vermont found itself in an atmospheric duct bearing moisture-laden winds with the raised terrain of the Green Mountains serving as a trap for the oceanic flow. As long as the southeast winds aloft prevailed, the precipitation process continued. Locations in western Massachusetts and Vermont received about double the amount of precipitation that might be expected from a normal coastal storm.

Not until the cyclonic center in the St. Lawrence Valley ended its northwest movement and veered to the northeast on March 5, did the cold front move eastward and wipe out the alignments that had sustained the southeast flow off the ocean. The three days of heavy precipitation tapered off on the 6th, though unstable air carried by westerly winds behind the cold front dropped additional light amounts in mountain-type snow showers.

171

A total of 50 inches of snow fell during this period in extreme southern Vermont at Readsboro at an elevation of 1,122 feet. Just over the border in Massachusetts, Peru, at 1,700 feet, measured 47 inches of new snowfall. These early March snowfalls, coming on top of a snowy January and February, raised snow cover to record depths. Readsboro reported 80 inches on the ground on March 5, a good part having fallen on the 2nd and 3rd when 2.55 inches of precipitation produced 23 inches of new snow in a twenty-four hour period.

Enormous amounts of moisture fell on the higher elevations of western New England. Two stations recorded over 4 inches water content of melted snowfall: Peru, Mass., 4.10 inches, and Readsboro, Vermont, 4.07 inches. Other stations along the Massachusetts-Vermont border exceeded 3 inches, and a sizable area in adjoining states had more than 2 inches of melted snow during the storm period.

The local amounts of snow accumulation were largely determined by temperature, with elevation above sea level proving the governing factor. Thermometers over the area hovered a little above or a little below the freezing mark. At higher elevations temperatures remained below 32° and all the precipitation fell as snow. Lower down, rain mixed with snow at times, and only rain fell on the valley floors. The readings at Adams in northwestern Massachusetts were probably representative of those in the all-snow area, ranging from 30° to 25° on the 3rd and from 32° to 22° on the 4th. Rutland was typical of valley locations with a span from 40° to 29° on the 3rd.

Highway traffic in all parts of the snow area experienced interruptions, and several small communities were isolated for two to four days by the deep drifts. Power and communication lines were downed by the wet, clinging snow.

The Great Mid-February Snowstorm of 1958

About once in a generation a great snowstorm comes along. On the Atlantic coastal plain in winter, "great" implies a storm possessing extreme physical force, large areal extent, and massive snow-producing abilities. These develop in the Gulf of Mexico or along the southeast Atlantic coast, and race northeast while spreading a mantle of white from the southern Appalachians to northern New England and beyond. Their impact over the entire region is remembered for many years, and their statistics hold a revered place in the annals of storm history.

No other snow event of the twentieth century can match the huge dimensions displayed by the Great Mid-February Storm of 1958, extending as it did along the entire Atlantic seaboard and inland to the Appalachians. In the nineteenth century, only three snowstorms rate as equals: The Great Snowstorm of January 1831, the Cold Storm of January 1857, and the Eastern Blizzard in February 1899. Even the Blizzard of '88 did not meet the above requirements since its geographical extent was limited to the Northeast.

The great snowstorm on February 15-17, 1958, spread a mantle of white to a depth of 12 inches or more in every state from Alabama to Maine. Across northeast Pennsylvania, central New York, and northern New England, its mighty stature was exhibited by deposits of snowfalls of 30 inches and more, which raised existing snow cover to depths unmatched in this century. New marks also were established for the concentrated rate of fall of the flakes.

Snowfall amounts over southern New England averaged about 20 inches. Several Vermont localities exceeded that figure, led by Cavendish with 24.3 inches. The descending flakes seemed to concentrate their greatest energy in the middle of the Connecticut River Valley where they piled up to a depth of 31.5 inches at Hanover and to 32.5 inches at nearby Lebanon Airport. Depths across the river in Vermont must have been similar. At the conclusion of the storm, the depth of new and old snow measured 60 inches at Lebanon and 45 inches at Hanover. In Vermont, Cavendish reported a depth of 53 inches on the ground and Montpelier Airport had 52 inches on the day following the storm.

Burlington totaled 15.9 inches on the two days, contributing the bulk of the record February snowfall of 34.3 inches. Since January had already broken the month's record with 33.7 inches, the two-month total set a new all-time mark for snowfall in two consecutive months.

It was a cold storm. Burlington's temperature on the 16th ranged from $6°$ to $0°$, and on the 17th from $7°$ to $-7°$. The thermometer slumped to a nadir of $-18°$ on the morning of the 18th. In the southeast, radiational cooling produced some surprisingly low readings, with Vernon dropping to $-33°$, the lowest in the state.

The storm assumed blizzard proportions over much of northern and western New England, especially in its later stages when the wind shifted to the northwest. The light-weight snow was piled into mountainous drifts by the gale-force winds, and the rapidly falling temperature caused a severe windchill and much discomfort. Drifts as high as 15 feet blocked roads in southern Vermont and the central Connecticut River Valley. Twenty-six deaths were attributed to the storm in the six New England states.

173

The End-of-Year Blizzard in December 1962

One of the severest blizzards in Vermont's history raged over all sections on the three closing days of December 1962. Though the actual snowfall amounted only to 2 to 11 inches in various sections, the gale-whipped flakes were carried along the ground in thick clouds, reducing visibility to near zero, and smothering attempts to maintain air, rail, and highway transportation. The below-zero cold on New Year's Eve was hardly bearable for outdoor activities and caused frostbite among those exposed for more than a few minutes.

The wind at Burlington peaked at 37 miles per hour from the north on the 31st when the temperature registered a maximum of 5° and a minimum of −10°. The arctic quality of the northerly airmass was illustrated by a reading atop Mt. Mansfield of −30° on the 30th and −40° on the 31st. These matched similar arctic conditions of −41° and −40° at the Mt. Washington Observatory on these days. The maximum reading at the top of Mt. Mansfield on the 31st reached only to a frigid −18°, giving a mean for the day of −29°. At West Burke in a valley location, the thermometer ranged from a high of −6° down to a low of −15° on the 31st. True blizzard conditions prevailed everywhere in Vermont.

All sections of the state reported trees and wires down, causing the loss of power for electricity and heat and disrupting phone services. Many windows were smashed by the high winds, TV antennas broken, and shingles ripped from roofs and sidings. The piercing cold sifted through cracks and crevices of buildings and froze water pipes that had never given trouble before during a cold wave.

Stores and offices in many communities remained closed all week since travel conditions continued very hazardous. Residents in remote areas were marooned for many hours or days behind snow-drifted roads. Ski operators closed down or greatly curtailed operations over the holiday as a result of the extreme wind-chill factor on the exposed slopes.

The weather map situation responsible for creating the blizzard conditions, although of infrequent occurrence, musters the heaviest weapons available in the arsenal of Vermont's winter weather. The center of action lay about two to three hundred miles east of northern Vermont in the Gulf of Maine where a storm stalled for a number of hours. The pinwheel circulation around the center of the deep barometric low pressure instituted a rush of arctic air borne on northerly winds from central Quebec and the Hudson Bay area directly into Vermont and the rest of New England.

When the coastal low-pressure center reached a position south of Cape Race, Nova Scotia, about noon on the 30th, it turned to the northwest

174

and performed a loop over the Bay of Fundy, into northeastern Maine, and back to Nova Scotia during the night hours of the 30-31st. The central pressure dropped to the low point of 28.73 inches at Eastport, Maine.

A warm easterly current moving around to the north of the low-pressure center clashed with the frigid air descending from central Canada. A distinct front between warm oceanic air and the cold continental air became established over eastern Quebec and interior Maine laying on a north-south axis. On the 31st, the maximum temperature at Caribou in northern Maine (in the warm airstream) stood at 29°, while at Bloomfield in northeastern Vermont, the cold airstream kept the thermometer at 2°. On January 1, Bloomfield rose only to −2°, but next day the warm air from the Atlantic Ocean, circulating around north of the low-pressure center, penetrated westward into Vermont, cut off the arctic air flow, and sent the thermometer soaring to 25° on the 2nd and to 30° on the 3rd at Bloomfield.

Railroad tracks as well as highways need to be kept clear of snow as is demonstrated here on the St. Johnsbury and Lamoille County Railroad. The plow is on the front of a "deadhead" locomotive which is being pushed (with a caboose in between) by two "live" locomotives. Photo courtesy of Morristown Photo Archive.

175

The Great Post-Christmas Storm of 1969

A storm center passing inside of Nantucket and Cape Cod over southeastern Massachusetts is likely to drop its heaviest snowfall west of the Green Mountains rather than east of the mountains in the Connecticut River Valley. When a storm of this type occurred on the three days following Christmas 1969, the northwestern corner of Vermont, along with adjacent portions of New York State and southern Quebec, experienced their greatest snowstorm of modern record. Burlington's total of 29.8 inches was the greatest ever to fall there in a single storm period, as was Montreal's 27.5 inches. Other Vermont locations reporting excessive amounts were Montpelier Airport with 34 inches and St. Albans Bay with 32 inches.

Early on Christmas morning Burlington's thermometer dipped to a low of $-16°$ after a night of intense radiation under clear skies, resulting from a large anticyclone located not far to the northwest in Quebec. Temperatures dropped as low as $-27°$ at Enosburg Falls and $-26°$ at Northfield under light wind conditions. Such a clear, still, cold morning carried little intimation that another twenty-four hours would bring on one of the mightiest snowstorms in all Vermont annals.

The weather map on Christmas Day placed a low-pressure system over Georgia and northwestern Florida, almost directly south of the high-pressure system overlying southern Quebec. The interaction between these two opposing pressure forces was to produce and then prolong the period of excessive precipitation in New England and Quebec during the next three days. As the low-pressure trough moved northeast and offshore on the 26th, it developed a series of wave disturbances along the front. The first depression arrived in the New York City area about noon, but then faded, and a second center about 250 miles to the southeast became the main scene of activity as it moved directly north to the vicinity of Nantucket and Cape Cod. On the morning of the 27th, the principal center was still in the vicinity of Massachusetts Bay, having remained almost stationary for about twenty hours. In the meantime, a third center developed on the front to the southeast and moved north well offshore, as the entire pressure trough gyrated counterclockwise from a north-south to an east-west axis. This initiated a vast flow of southeast winds bearing huge amounts of moisture from the surface of the Atlantic Ocean to New England, over the heights of the Green Mountains and beyond.

The landward flow of warm oceanic air brought above-freezing surface temperatures over the coastal plain of Maine and New Hampshire and well into the interior. During the period of heaviest precipitation, the surface freezing line ran through western New Hampshire, north and south, almost parallel with the Connecticut River.

The type of precipitation changed as one progressed from east to west.

176

Rain fell along the coast, a mixture of rain and frozen ice pellets (sleet) in central New Hampshire, freezing rain and light snow in the Connecticut River Valley, and heavy snow along the eastern slopes and on the crest of the Green Mountains. The Champlain Valley also received heavy snow as the temperature at both the surface and aloft remained well below freezing.

The first flakes at Burlington began to filter down soon after midnight, at 12:44 A.M. on the 26th. They fell lightly through the early morning hours, but after sunrise increased. By 9:30 A.M. the light snow increased to moderate intensity, resulting in an accumulation of five inches of new snow by 1:00 P.M. The thermometer rose steadily from 9° at midnight to an afternoon high of 22°, and then remained close to 20° during most of the afternoon. Despite the low temperatures, some sleet fell briefly during the mid-afternoon, illustrating the complex structure of the air strata overhead. Another 3.2 inches of new snow fell during the afternoon. The day's total was 8.2 inches.

The snowfall tapered off during the night with only insignificant amounts descending. It looked as though Burlington was to experience only an average snowstorm for the season, but changes in process in the weather situation along the New England coast soon affected the Burlington area. Another period of moderate snow commenced about 5:30 A.M. on the 27th, and at 7:00 A.M. another 5.6 inches of new snow lay on the ground. The flakes continued to filter down throughout the 27th at a rate of about a half-inch per hour, increasing the snow depth by 6.8 inches during the daylight hours. Temperatures ranged from 15° to 21°, close to the previous day's spread. The total for the twenty-four hours was 12.4 inches.

The snowfall continued at about the same rate overnight, mostly light, though with one period of moderate intensity before 6:00 A.M. of the 28th. The overnight hours brought 6.7 inches to swell the grand total for the storm to 29.7 inches. The snowfall finally ceased at 3:25 P.M. on the afternoon of the 28th after sixty-one hours and ten minutes of continuous activity. The water content of the snow equaled 2.73 inches, a very substantial total for a wintertime storm. Other stations measured water content from 3 inches to as high as 4.78 inches at Rochester.

Across the northwestern portion of the state new snowfall totaled from 18 to 36 inches with a local high of 45 inches reported at Waitsfield. This was the greatest single snowfall total since the March storm in 1947 produced great accumulations at the high elevations in the south.

Since these mighty snowfall amounts closely followed a major storm on December 22-23, snow depths approached record levels. Drifts commonly mounted to six feet and in some places up to 30 feet, closing roads and halting virtually all transportation except by snowmobile. The state was declared a disaster area by Governor Deane C. Davis so that federal assistance could be obtained for the task of digging out. Coming at the peak of the holiday skiing season, it caused a second storm, a howl of controversy between the Governor and those engaged in the winter ski industry who feared their customers would be deterred from venturing to Vermont by the disaster declaration.

Four Snowy Winters: 1969 to 1972

Remember the popular bumper stickers: "I survived the Winter of '69?" Little did Vermonters realize that the same slogan would apply for the next three years also.

The four winters from 1968-69 to 1971-72 make up the snowiest series in the entire recorded meteorological history of the state dating back a century and a half. Mention has been made of the anomaly of snowy winters tending to come in pairs; now the present generation has seen two pairs, back to back!

The snowfall figures at Burlington Airport demonstrate the unique series. The mean snowfall for a thirty-year period is 79.1 inches, but the Queen City had four successive winters with 96.3, 104.6, 145.4, and 108.9 inches, or in percentage of normal, 122, 132, 184, and 138.

Aside from the 111.6 inches in 1965-66, one must go back to 1887-88 for a season at Burlington with more than 100 inches of snow, and then there were two such in a row: 1886-87 with 132 inches and 1887-88 with 113.5 inches. Burlington has a relatively small annual snowfall as a result of its low elevation and distance from moisture sources.

Mt. Mansfield is the snowiest weather station currently operating in Vermont since its elevation is about twice that of other reporting localities. The seasonal figures in inches at high elevations for the four winters under survey were:

	Mt. Mansfield 3,950 ft.	Waitsfield 820 ft.	Peru 1,670 ft.	Montpelier 1,126 ft.
1968-69	327.0	176.0	161.0	127.8
1969-70	255.4	151.5	136.0	127.9
1970-71	318.6	198.5	184.0	139.4
1971-72	266.0	173.0	170.6	127.8

The Big Storm on February 6-7, 1978

The biggest snowstorm of the late 1970s hit some parts of New England hard and let others off with only a light visitation of the white flakes. Vermont shared this pattern, with a strip across the central part of the

178

state buried under 20 inches and more while the north and south received much less.

The storm center stalled for many hours on February 6-7, 1978, off the tip of Long Island, all the while producing a snowfall of thirty hours duration over the land to the north. At many points in southeast New England the heaviest single storm of record resulted when nearly 30 inches fell at Providence and Boston.

The storm system possessed an unusual organization, being extended west-to-east south of Long Island for many hours without apparent movement. To the north of the low pressure axis, the winds had a long trajectory over the open Atlantic Ocean, a situation similar to the March Blizzard of 1888 and also to the big snowstorm on December 26-28, 1969.

The zone of 20 inches in Vermont included: Cavendish 25 inches, Rutland 24 inches, Vernon 22 inches, Union Village 21 inches, and Peru 20 inches. The contrast of 24 inches at Rutland and 3.9 inches at Burlington was remarkable, as was the valley location of Vernon with 22 inches and lofty Readsboro with only 10.8 inches.

In the Connecticut Valley of New Hampshire, the following totals were reported: Lancaster 24 inches, Hanover 25 inches, Keene 20 inches, and Walpole 15 inches, yet only seven inches at First Connecticut Lake in the far north.

May Snowstorms of the
Nineteenth Century

Backward springs can cause much comment and discouragement among farmers and gardeners, and the view of wet flakes falling outside in late spring brings dismay to all who anticipate the pleasures of summer. Two backward springs achieved notoriety during the days of pioneer settlement in the eighteenth century. During the "scarce season" or "starving time" in 1789, a snowstorm was reported on May 8 by the Reverend Nathan Perkins at Tinmouth in Rutland County, and the mountains were white with snow that year as late as May 20. Just a decade later, during another spring noted for cold and frost, 12 inches of snow fell on May 5 in the upper Connecticut River Valley at Lancaster, New Hampshire, opposite Lunenburg, and snow banks were still to be seen around Hardwick on June 9.

A snowstorm of wide extent covered much of the Northeast on May 8, 1803, bringing approximately six inches to southern New England locations close to the track of the storm. The canopy of snow reached as far north as Windham County in Vermont and Cheshire County in New Hampshire. People in Keene, New Hampshire, went to church in sleighs on that May Sabbath, so the accumulations must have been several inches.

The cold period in early May drew the attention of the *Vermont Gazette* at Bennington:

> On Friday [6th] it was cold and raining a greater part of the day, and in the night there was a severe frost for the season; on Saturday it was so cold as scarcely to thaw during the day; on Sunday it snowed considerably a great part of the day, and yesterday continued uncomfortably cold, notwithstanding the sun shone fair. Our expectations with respect to fruit cut off, as even the young currants on our bushes appear to have been killed by the frost.

The "cold years" from 1812 to 1817, in addition to the famous June snow of 1816, were marked by a series of May snowstorms: on May 4, 1812, May 3, 1814, May 19, 1815, May 14-15, 1816, and May 15 and 27, 1817.

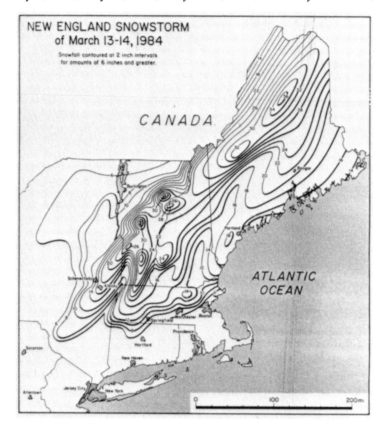

HEAVY SNOW IN NEW ENGLAND ON MARCH 13-14, 1984

On March 13th and 14th a strong winter storm moved up the east coast and dumped up to 3 feet of snow on New England. Heaviest amounts were recorded at the higher elevations in Vermont, New Hampshire and Maine. Telos Lake, Maine, about 100 miles north of Bangor, received the maximum amount of 36 inches. — Data supplied by NWSFOs at Portland, Maine; Boston, Massachusetts; and Albany, New York.

THE VERMONT WEATHER BOOK

The Great Snowstorm of Mid-May 1834

The greatest of all May snowstorms to affect the Northeast in recorded meteorological history occurred in mid-May 1834. A light snow began in Ohio early on the fourteenth and spread across northern Pennsylvania and New York State. The intensity increased to moderate in central Vermont and to heavy and excessive in northeast Vermont and northern New Hampshire. At the valley location of Rutland, the snowfall started at 8:00 P.M. on the fourteenth and continued through the afternoon of the next day, with 12 inches accumulating. Burlington also received the same amount in "a tremendous storm," the heaviest late-season storm since 1816. Over the crest of the Green Mountains in east-central Vermont, similar amounts covered the landscape. Woodstock had four to five inches downtown, with an estimated foot on the surrounding hills, and at Randolph, weather-watching William Nutting noted that the twelve-hour storm "fell near one foot." The observer at Hanover, New Hampshire, reported a fall of 15 inches.

Northward in the upper Connecticut River Valley, temperature conditions and the supply of moisture proved ideal for heavy snow making, especially on high ground where the thermometer stood close to the freezing mark.

Marshfield, in Washington County northeast of Montpelier, received a blanket "more than two feet deep," according to its historian. At Newbury, David Johnson reported the snow depth at two feet also, enabling sleighs to put in an appearance in mid-May. Across the Connecticut River at Haverhill, New Hampshire, two feet of snow descended in the valleys and three feet on high ground. The stage from Danville to Montpelier became stuck in the deep accumulations when trying to climb a hill and had to be abandoned.

Jonas Tucker of West Newbury entered in his account book:

> May 15: Snow fell from daylight to ten o'clock two feet deep, on higher land it fell two and a half—had it not settled it would probably have been from two and a half to three and a half deep. For about two hours it gained an inch each ten minutes.
> May 16. The earth was completely covered with snow all day.
> May 17. Bare spots appeared.

May snowstorms continued to whiten the countryside during the remainder of this cold period of the early nineteenth century, when the Little Ice Age was in full swing. On May 13, 1836, both Zadock Thompson at North Hatley, Quebec, and James Johns at Huntington noted snowfalls. On the first days of May 1841 William Nutting recorded two snows at Randolph: eight inches on the 1st and four inches on the 2nd. The backward season of 1843 was capped by snowflakes in the air on May 31 and June 1 throughout the Green Mountain region. David Johnson at Newbury observed three inches on the surrounding high ground on the last day of May in 1843.

During the cold period spanning the middle years of the nineteenth century, snow fell in May, enough to lie on the ground, that is, in every year from 1854 to 1858, in 1860, 1862, 1863, (1864 had a light snow in June), 1865, 1866, 1868, 1869, 1870, 1872, 1873, 1874, 1875, 1877, 1878, 1882, and 1884. Then there appears to have been a skip until the big snow of May 1892.

The Decoration Day (May 30) Snowstorm in 1884

At Randolph: "On the 28th, 29th, and 30th, there were hard freezes, and on the 30th there was a snowstorm that will be long remembered, as the flowers that decorated the soldiers' graves were covered with snowflakes" ("R" in *Country Gentleman*, June 12, 1884).

At Lunenburg: Hiram Cutting measured two inches of snow on the 30th. His low temperature was 35°.

At Windsor: the last three days of May saw the mercury at 32° each morning: "Although the freeze caused some forest trees to turn red like autumn, it is hoped the apple crop is not damaged," wrote the local observer. ("J. L. M." in *Country Gentleman*, June 12, 1884).

The Great Late-May Snowstorm in 1892

"The snowstorm of the 20-21st [was] the greatest in the memory of the oldest inhabitants here so near the close of the month; 55 years [actually 58] years ago [on] the 14-15th the snow fell two feet. So wrote H. F. J. Schribner of Strafford, an observer for the New England Meteorological Society. He added that snow fell from 3:30 A.M. to 11:00 A.M. on May 20, 1892; changed to rain at noon and continued into the late night; snow mixed with the rain descended from 5:00 A.M. on the 21st to noon; then snow took over and fell heavily until late in the second night. He measured 10 inches of snow on the 20th, and six inches on the 21st. His thermometer hovered just above the freezing mark, from 33° to 35° during the snow periods. On higher ground where the temperature was a degree or two lower, even more snow piled up. One report to the society spoke of 28 inches of wet pack. Total precipitation measurements across Vermont ranged up to 2.63 inches at Hartland, confirming that snowfall amounts of 28 inches were possible in this storm.

June Snows

"What is so rare as a snowstorm in June?" Vermont has experienced two substantial snowfalls in the month of the summer solstice, and on three other occasions light accumulations covering the ground have been reported. The latter occurred on June 10-11, 1773, at Newbury; on June 9, 1833, over most of the state, and on May 31-June 1, 1843, across a greater part of the north. All happened prior to 1850 during the period of the Little Ice Age when temperatures in the Green Mountains were perhaps two to three degrees lower than during the first half of the twentieth century.

Snow flurries have been spotted in some recent Junes, and occasionally some of the higher peaks have worn a snowy crown for a few hours, but nothing of the nature of the occurrences in June 1816 and June 1842, when deep snow covered valley locations. (The snow event in June 1816 is discussed in the chapter on the summer of that year.)

June 1842

A general snowstorm covered most of the Northeast on June 10 and 11, 1842, with amounts varying from a trace in Pennsylvania and Connecticut to as much as 10 to 12 inches in northern Vermont and New Hampshire. Snow was noticed at Western Reserve College at Hudson, Ohio, on the late evening of the 10th, and next day it became general in the east. James Johns, as usual, took note of the occasion at Huntington:

> June 11. A memorable day in the calendar. A snow storm commenced about daylight and continued with more or less severity until about one o'clock P.M. when it ceased. The clouds dispersed and the sun came out pleasant and the ground which was white with snow was soon bare again.

Zadock Thompson at his station on the University of Vermont campus noticed the snowfall and gave some meteorological details about June 11: "Snow during the whole forenoon – boards whitened and the mountains white as in winter." His thermometer read 34° at sunrise and 40° in early afternoon. The north wind prevailing at Burlington and at Plattsburgh, New York, across the lake would indicate the passage of a low-pressure center to the southward, hence the cold air flow and the frozen flakes. A dispatch from Vergennes stated that in hill towns 10 miles to the east the snow lay six inches deep.

In eastern Vermont, snow fell at Randolph from before sunrise until afternoon. At Hanover, New Hampshire, a mixed rain and snow was experienced during the day until 4:00 P.M.; the total precipitation when melted was 0.40 inch.

Farther north the depths increased. A press dispatch from Orleans County reported 10 to 12 inches at Irasburg and Barton. At the same latitude in Coos County, New Hampshire, a local historian of Berlin mentioned a depth of 11 inches. An editor there exclaimed: "This beats the year 1816 and all others within our memories."

Snow in July and August

No direct evidence has been found of an authentic snowfall at valley locations in Vermont during the summer months of July and August, not even in the "Year Without a Summer" in 1816, though mountain peaks have been whitened on several occasions. Thunderstorms attending cold fronts were the probable source of below-freezing temperatures aloft that cause the frozen precipitation covering the highest peaks while rain showers fall in the valleys below.

Only two falls of snow in local areas have been reported in summertime. In July, 1888, a surprise snowstorm caused a group of mountain climbers to remain atop Camel's Hump overnight, and in August 1954, heavy snow fell at two Vermont communities for periods of about an hour.

July 1888

The Burlington *Free Press* carried the following dispatch from Waterbury, dated July 12, 1888: "A terrific westerly wind and rain storm prevailed last night and today. Trees were blown down and much hay was damaged. A company of 12 young ladies and gentlemen on top of Camel's Hump in a tent last night suffered from a snow storm, accompanied by wind and rain. Help with teams rescued them today."

These conditions resulted from a severe cyclonic storm that swept over New England on July 11-12, 1888. It was particularly damaging across northern New York State and Vermont where it took on the characteristics of a straight-line "wind-rush" or "downburst" during a thunderstorm, downing and uprooting trees over a lengthy strip of land in a direct line.

August 1954

The U.S. Weather Bureau's *Climatological Data* for August 1954 mentions a number of unusual weather events occurring in New England during the month. One of the assortment took place on the 11th and was described: "heavy snowfall for about an hour in two Vermont communities—Beecher Falls and Lincoln—this is probably the earliest date of such an occurrence in New England."

184

Some unusual physical mechanism would be required to create snowfall at ground level under the existing temperature conditions. The lowest temperature reported in New England on this date was 35° atop Mt. Washington; and in Vermont, 48° at West Burke and Cornwall were the minimums, and these probably occurred close to sunrise. Most likely an atmospheric "down-burst" of very cold air from aloft was responsible. These occur when a body of air in a turbulent cloud is carried down to ground level. If the general freezing level is at a relatively low elevation, the descending air might remain below the freezing point throughout its descent, despite the normal dynamic heating that takes place in a descending air current.

Since the two communities of Beecher Falls and Lincoln (described below) are not adjacent, separate thunderstorms occurring along an extensive cold front must have given rise to the disparate snowfalls reported. Beecher Falls is located in the extreme northeast corner of the state on the Connecticut River, while Lincoln is a hill town in eastern Addison County on the western slope of the Green Mountains.

Talk about the old days! King Winter got off to an early start as he ignored the calendar to decorate two Vermont communities with a mantle of honest-to-goodness snow in August.

It may be a little early to dust off the red flannels, but folks in Beecher Falls and Lincoln are wondering what will happen next. They've got nothing on the weatherman.

It actually snowed for almost an hour in Lincoln on Wednesday. Daniel L. Garland, veteran town representative, termed it the "worst snowstorm" he had ever seen.

He said it snowed so hard automobile drivers had to stop their cars. The white stuff blanketed the entire area and there were still some patches of snow on the ground as late as noon yesterday.

The snow puffed in between two thunderstorms as the temperature dropped to 50. The valley caught more snow than the hills, and gardens and shrubbery were ruined.

Mrs. Garland reported "bushels" of fallen leaves beneath maple trees. No one in the community could remember a snowstorm in August before this.

Beecher Falls, tucked up against the edge of Canada in the northeast tip of Vermont, reported a similar freakish snowfall on Wednesday.

Again the snow followed on the heels of a violent thunderstorm and when the snowfall had passed, enough snow remained on the ground to provide ammunition for some first-class snowball fights by the enterprising small fry of the community.

Burlington Free Press, August 12, 1954.

September Snows

Three noteworthy snowstorms occurred in September during the nine-teenth century when the ground in some parts of the state was covered to the substantial depth of six inches or more, in 1835, 1844, and 1885. An early snow squall in September 1812 should be mentioned, though the ground was merely whitened without much accumulation. In harmony with the general warming trend of recent decades, no substantial snowfalls have been recorded in the month of September during the present century.

September 1835

Snow commenced to fall at the early date of September 30, 1835, across northern Vermont, northern New Hampshire, and adjacent parts of the Province of Quebec. Professor Zadock Thompson of Burlington, then residing at Hatley, Quebec, about 25 miles north of the Vermont border, made the following pertinent entry in his weather register for September 30: "Began to snow this afternoon at Hatley - snowed all night and all the next day [Oct. 1], in which time the snow fell 10 inches deep on the level. Although it rained on the 3d of October, the snow lasted more than a week."

The snow canopy extended southward into central Vermont and New Hampshire as the weather observer at Dartmouth College at Hanover, New Hampshire, noticed snowflakes in the air on the 30th and his ther-mometer down to 40°. Farther up the valley, as much as six inches fell at Kilkenny near Lancaster, New Hampshire. Across the mountains in northwestern Vermont, the local press reported depths from 6 to 12 inches in Franklin County where lake-effect snows usually add to the total amount. Trees in green leaf gathered heavy amounts of the moist snow; many were destroyed either by breaking their trunks in two, splintering, or being torn out by the roots.

James Johns at Huntington noticed snow on the top of Camel's Hump on September 30; next day his diary entry read: "Cloudy, cold toward night it began to snow light flurries mixed with rain which increased as the day shut in and the ground white after dark." The following morning the sun took all the snow off.

September Snow in 1844

A most unusual early-season snowstorm for depth and extent of in-fluence swept across the Northeast on the next to last day of September in 1844. In northwestern Pennsylvania in the Bradford hills and in the elevated plateau of western New York as much as 8 inches were reported. The snow belt extended eastward across central and northern Vermont.

A snowy day occurred at Washington Academy, Salem, New York, on the immediate Vermont border, and William Nutting at Randolph in the central part of the Green Mountain State measured two inches on the ground: "Sept. 29—began to snow fast at 8½ A.M.—snow and rain all day—snowed fast in evening—cleared off in the night—fell two inches lying on the ground." Another weather watcher at Wheelock in northern Caledonia County noted six inches on the 29th. A press report from Stewartstown, New Hampshire, across the river from Canaan, Vermont, on the Canadian border told of eight to nine inches and good sleighing after the storm. The snow shield apparently did not extend into the valley locations of southern Vermont nor the Champlain Valley in the vicinity of Burlington. Professor Zadock Thompson recorded no snow, but his thermometer went down to 28° on the 28th with a hard frost covering the land and a heavy fog blanketing the lake in a white cloud.

The September Snow in 1885

The heaviest September snowstorm since 1835 spread over much of northern Vermont on the 22nd and 23rd in 1885. Atop Mt. Mansfield, observer William Barney reported a fall of 12 inches with the temperature at 37.5° on the 23rd and 31.5° on the 24th.

Barney wrote: "Snow disappeared a few days after it fell although drifts of three feet were formed in many places. Earliest snow anybody had ever seen here. Extended over a greater part of the state, also in New Hampshire and Canada. Could not see it went into New York State. Snow flying the first week of September."

The storm seemed to center its fury in Orange County. H. F. J. Schribner at Strafford reported: "Snow commenced to fall at 7:00 A.M. (23d) continued to 11 P.M. 9 inches falling. 12 miles north of me 12 inches fell. The damage to corn . . . . fruit & the forest was great, looking in some places as if a cyclone had passed through. It remained two days so farmers had to fodder as in winter." His three thermometer readings on the 23rd were all 38°, so it must have been a very damp snow.

At Post Mills, a 472-foot elevation also in Orange County, the observer recorded: "A snowstorm visited this place on the 23 inst., lasting from 7:30 A.M. to 4 P.M. and about two inches being the most noted. On the hills in the vicinity of Strafford, Vermont, the snow fell to about eight in. deep and drifted in some places about three feet deep." His thermometer read 36° at 2:00 P.M. At Dorset in the southwest only a trace was reported, but the observer noted it was four inches deep on the mountain six miles to the east. His thermometer read 35.5°.

September Snows in the Twentieth Century

Several Vermont locations at high elevations and different exposures have experienced a rare September snowfall. A light blanket of white covered the floor of the Connecticut Valley at the end of September, 1911, as far south as Norwich, Vermont, and Hanover, New Hampshire, with two inches reported at the latter.

A semblance of a September snowstorm occurred on September 24-25, 1950, when Chelsea and First Connecticut Lake, New Hampshire, reported measurable amounts, though trifles: 0.6 inch at the former and one inch at the latter. Mt. Washington had 2.5 inches. Traces fell at several Vermont communities across the north, including Burlington.

Burlington has witnessed six traces of snow falling in September in this century, all in the years from 1925 to the present, but no measurable amounts have been measured at the Queen City. The most recent occurrence of September snowflakes at Burlington came on September 22, 1963.

October Snows

October snows appear to have been unusually frequent during the first four decades of the settlement of Vermont, storms being noted in Vermont or in adjoining territory in the Octobers of 1764, 1765, 1769, 1775, 1777, 1780, 1783, 1788, 1789, and 1791. The series of early-season snowfalls supplies further evidence of the lower level of temperatures prevailing during the Little Ice Age in America.

1946

The closest approach to a general September snowfall over a wide area in this century occurred just a day late to qualify, on October 1-2, 1946, when as much as 12 inches fell at Searsburg Mountain (2,360 feet) and 0.5 inch at Rutland in the valley. St. Albans in the northwest corner of the state reported three inches and locations across Lake Champlain in northern New York State had six inches.

The Hurricane Snowstorm in October 1804

The greatest October snowstorm in Vermont history was caused by a tropical storm from southerly latitudes. The track of the storm center, which raised hurricane-force winds across southern New England, cut through the interior of the Middle Atlantic States and then across southern New England approximately on a Danbury-Hartford-Boston line. Heavy rains fell over southern Connecticut, Rhode Island, and eastern Massachusetts, but north of the storm track in the hill country cold Canadian air entering the circulation clashed with the tropical air brought north by the hurricane. Temperatures fell low enough on Tuesday morning, October 9, 1804, to produce frozen precipitation over the northern half of interior New England. Amounts were very heavy since the moisture-laden air from the Atlantic Ocean was forced to rise over the elevated ground of New Hampshire and Vermont. The wet snowflakes clung to the trees whose branches were still in leaf, doing enormous damage. The extensive tree losses were recalled thirty-two years later by Benjamin Harwood of Bennington when another similar storm occurred in October 1836 (though its damage was less severe).

"The Remarkable Snow Storm" was reported in the local press of the Connecticut River Valley:

> On Tuesday last (9th instant) about the middle of the forenoon, the weather suddenly changed from temperate rain, to a storm of snow, attended with thunder and lightning and violent wind on the high lands—the storm continued with some intermissions till Wednesday morning. It is judged that the mean depth which fell was from 15 to 18 inches. Contiguous to the river, it dissolved rapidly, yet repeatedly measured from 4 to 5 inches, and covered the ground for more than 30 hours. On the hills it was blown considerably into drifts, which in places covered the fences and blocked up the roads.
>
> *Political Observatory* (Walpole, N.H.), Oct. 13, 1804.

1836

The coldest year of the nineteenth century in the Northeast was 1836, and three widespread falls of snow during the last days of September and the first two weeks of October provided the highlights of an early fall season. Our weather-conscious diarist at Bennington, Benjamin Harwood, and the weather observer at nearby Williamstown, Massachusetts, provide first-hand accounts of the unusual events.

Harwood reported "rain and snow intermingled" on the morning of September 26, and the arrival of a strong cold front on the 27th with northwest winds and "a kind of tornado which however was quickly over." His thermometer read 40°. Snow fell again on the morning of the 28th, and the mountains around Williamstown (Mass.) were white. Harwood's thermometer stood at 34° early on the 29th.

Just a week later, on October 5, a heavy snowfall was reported in central New York State—as much as two feet covered the high ground in Onondaga County. At Bennington in Vermont, Harwood described the storm in his vicinity: "there was snow mingled with rain—Mountains white—36°." On the 7th, he noted: "Manchester Mountain very white."

On the third Wednesday in a row, October 12, snow fell on the high ground of Vermont and western Massachusetts, and this time reached to the valley floors in many places. Early risers at Williamstown, Massachusetts, on the 13th, were greeted by a covering of three inches and at Hanover, New Hampshire, it accumulated to five inches before turning to rain about ten o'clock in the morning. Harwood's entries on the 12th and 13th supply the weather details about the valley village of Bennington:

> October 12—Long to be remembered on account of the distressing snow storm which occurred at this date—Many trees of the forest retained green leaves, but the snow did not fall as heavy nor did it in any way operate as injuriously as on Oct. 9, 1804—Thermometer indicated—morning 33°, noon 42°, evening 33°.
> October 13—Every appearance of winter—ground covered with 4 to 6 inches depth with snow—SW wind blew and carried off most of it—Roads muddy, 33°, 48°, 50°.

The Snowiest October—1843

The snowy season of 1843-44 was introduced by a major snowstorm on October 22-23 that covered both valley and mountain locations with a deep mantle of 12 inches or more, and the snow cover was said to have remained until spring on high ground. Both Barnard and Newbury reported 12 inches on this occasion, as did Haverhill, New Hampshire, in the Connecticut River Valley. Press dispatches mentioned as much as two feet at unspecified locations and drifts to four feet. Hanover, New Hampshire, had six inches as indicative of the cover in the middle Connecticut River Valley. Snow apparently fell over the entire state, being reported as far south as Troy, New York, and Northampton, Massachusetts.

At Burlington, Professor Zadock Thompson measured 6 inches on the 23rd. His thermometer hovered near 32° during the storm. Another four inches came on the 27th with the mercury down to 28°. On the last day of the month, the ground was "mostly covered" and some snow remained through November 12 in the Champlain Valley. At Huntington high on the slopes of Camel's Hump, a wet snow of six inches broke some branches on trees, but by October 30 it turned "thawy to leave ground in some measure bare," according to James Johns.

October Snows of the Twentieth Century

During the present century, October snows of measurable depth have fallen at valley locations in 1906, 1907, 1910, 1917, 1918, 1925, 1926, 1933, 1934, 1952, 1965, 1969, and 1979.

The most notable of the early storms of this century came at the height of the foliage season on October 10, 1925, on a Saturday when many autumnal events were scheduled. At valley locations, depths up to 12 inches were reported, while on higher elevations amounts ran as high as 18 inches. High winds accompanying the storm caused considerable drifting. In the absence of rural plowing, many motorists became stranded

190

on mountain roads. Three miles west of Newport a drift five feet deep blocked the road to Montpelier. Smugglers Notch was closed by drifts from six to eight feet deep, and in the Stowe area some cars were snowed in until the following spring.

Burlington measured a fall of 4.4 inches with a wind of 36 miles per hour creating drifts. Snow plows were called out to clear the city streets.

The following year brought another October snowfall, though the storm conditions were not as severe. Amounts ranged up to eight inches at Danville and seven inches at Northfield on the 20th and 21st in 1926. The damp flakes clung to trees and wires; a large number of telephone poles toppled under the weight and communications were disrupted for several days. St. Albans was cut off from the outside world for many hours. All stations in Vermont reported three inches or more, and the state averaged 5.2 inches.

COLD WAVES

To a Vermonter, a thermometer isn't worth the liquid in its tube unless it goes down to $-40°$. On annual visits to the Green Mountains as summer residents, we always liked to listen to our neighboring farmers revel in tales of their $-40°$ thermometer readings during the winter. "It went down to 40 below and stayed there for days," was a common comment on the severity of the last winter.

It was not until much later in life that I had access to the historical weather records for Vermont and was able to check on some of those blood-chilling tales. I found that over the period of this century, from 1901 to 1984, the official thermometers of the national weather service, exposed under standard conditions, had fallen to $-40°$ or below in only eight Januaries, in only five Februaries, and in only two Decembers during the eighty-four winters. So in the 270 months involved, readings of $-40°$ occurred in only fifteen of those months, equal to 5.5 percent of the winter months - a frequency of one month every five and a half winters.

On the thermometer's scale, $-40°$ is an important figure because it is close to the lowest reading any area in Vermont can experience and it is also the point where the Fahrenheit and Celsius scales agree ($-40°F$ equals $-40°C$, with no fractions involved). If the thermometer always stood at $-40°$ (God forbid), there would be none of those painful mental calculations to change Fahrenheit into Celsius, or vice versa. Also it is very close to the freezing point of mercury, or, if you will, the melting point: $-37.97°F$ equals $-38.87°C$.

THE VERMONT WEATHER BOOK

Cold Friday in 1810

No other single weather day has been enshrined in the folklore of New England as often as "Cold Friday," which occurred unexpectedly on January 19, 1810, in the middle of one of the mildest winters of record. A century ago, whenever a northwest gale roared outside and the thermometer took a plunge, the story of the bitter gale and its tragic consequences was told and retold around the fireplaces of New England.

Though no thermometer reading for the day in Vermont has survived, comparative data from the New Hampshire side of the Connecticut River Valley indicated the probable degree of severity in Vermont. At Marlborough near Keene, New Hampshire, a thermometer stood at 51° at 11 A.M. on January 18, but in only twenty-four hours plunged to −12°, a descent of 63 degrees. Farther north in the central part of the upper valley at Hanover, the *Dartmouth Gazette* stated: "on the morning of the famous *cold Friday* the mercury stood at 17 or 18 below 0, at this place." Northward in the Connecticut River Valley at Lancaster, New Hampshire, across the river from Lunenburg, schoolmaster Adino Brackett left an informative diary entry:

> The weather until Friday the 19 inst. has been very changeable but at no time remarkably cold since the commencement of winter. We have not had a week without thawy days. But on Friday last [19th] the weather changed in a very sudden manner, and became remarkably cold and severe. One half of the scholars who attended my school were frozen, and this severity of weather still continues. At noon today Seth Willard dismissed his congregation on account of the severity of the cold. I am much deceived if I ever experienced colder weather.

Cold Friday was made doubly memorable by the tragic events which took place at Sanbornton in south-central New Hampshire. The Ellsworth family home was demolished during the early morning hours by the force of the gale raging across the open fields. The bedding and clothing of the family were blown in helter-skelter fashion across the farmland. In attempting to reach the shelter of a neighbor's house, the family sleigh was upset and damaged, and the occupants thrown into the snow. While the parents were seeking another conveyance at the neighbors, their three small children froze to death. Both parents carried the effects of their ordeal for their lifetimes. It served for all as a warning of the dangers of exposure to bitter windchill.

The Cold Week in 1835

The first full week in January 1835 became known as "The Cold Week" throughout New England for unprecedented low temperatures and their continuance for seven days. The thermometer on the Dartmouth campus at Hanover regularly received attention three times a day: at 7:00 A.M., 2:00 P.M., and 9:00 P.M. The mercury went below zero at the morning observation on every day from January 1 through 12. The coldest day was the 4th when the three readings were: $-32°$, $-7°$, and $-20°$, for a mean of $-19.7°$. Other minimums were $-23°$ on the 8th and $-26°$ on the 5th. No other week to that time was known to have sustained this degree of cold for so long.

Other thermometers over northern New England dropped to similar depths. In some cases the mercury froze, indicating a mark of $-40°$ or less had been reached. If spirit thermometers (containing alcohol, which does not freeze at $-40°$) had been available with a registering device, no doubt the magic $-40°$ figure would have been surpassed.

Niles' National Register of Baltimore, Maryland, the news weekly of its day, took notice of the extreme cold by carrying a summary of temperature reports that appeared in various local newspapers. Some of the Vermont figures attained either on January 4 or 5 of 1835 follow:

Montpelier $-40°$	Newbury $-36°$	Rutland $-30°$
White River $-40°$	Norwich $-36°$	Burlington $-26°$
Bradford $-38°$	Hyde Park $-36°$	

The Cold End of January in 1844

The last week in January 1844 proved outstanding for the long duration of severe cold. The low thermometer readings from January 25 to 31 were cited as an example of how cold it could get in Vermont by Dr. Hiram Cutting of Lunenburg, the state's leading meteorological authority in the post-Civil War years.

E. S. Paine of Randolph frequently reported his thermometer readings to the Boston and New York press, apparently taking pride in the coldness of his surroundings. In a letter to *The New York Weekly Tribune*, he reported the following minimum readings from January 25 to February 1, 1844: $-6°$, $-26°$, $-36°$, $-42°$, $-34°$, $-32°$, $-35°$, and $-32°$. No doubt he lived in a cold hollow, of which Vermont abounds, and wanted everyone to know about its frigidity.

Similar readings were reported to the *Boston Daily Advertiser* from Northfield, with a nadir of $-42°$ on the 28th. The thermometer at Dartmouth College produced a mean of its 7:30 A.M. readings of $-24.7°$ for

In the coldest winters river ice sometimes backs up, overflows the river banks, and blocks adjacent roads as in this photograph by Everett Vaile showing an ice back up on the West River which covered the road between Londonderry and South Londonderry. A one lane path had to be cut through the ice to allow travel between the villages.

the seven days following January 25. The lowest on Professor Charles Young's thermometer was −30°, and according to him, this last week in January was the coldest string of days from 1835 to 1857. Professor Thompson at Burlington also had seven consecutive days with minimums below zero, but his figures were not as extreme as those registered at Hanover.

The coldness of this period over the entire New England area is given graphic representation by a Courier and Ives print showing the early Cunardliner *S.S. Britannia* leaving Boston harbor by way of a seven-mile channel cut through the ice by port authorities.

Cold Temperatures in the Middle Connecticut River Valley: 1835-1857

An interesting list of cold periods at Hanover, New Hampshire, appeared in an 1857 issue of the *American Journal of Science*, published at

New Haven, Connecticut. The contributor was "Mr. Y," no doubt Professor Charles Young, the astronomer at Dartmouth College:

	Commencing	7:30 A.M. Mean	Minimum
7 days	Jan. 25, 1844	−24.7°	−30°
6 "	Feb. 2, 1836	−20.8°	−29°
10 "	Jan. 3, 1835	−15.6°	−32°
8 "	Feb. 13, 1849	−15.4°	−21°
7 "	Jan. 16, 1852	−11.8°	−23°
4 "	Jan. 22, 1857	−25.8°	−39°

On December 16, 1835, the thermometer read −17° at noon and the cold was accompanied by a strong northwest wind.

Cold Friday II: January 23, 1857

January 1857 stands preeminent in the weather annals of the nineteenth century as the coldest month of that century, second only to January 1780 as the coldest ever experienced.

Two severe cold waves came back-to-back after midmonth of January 1857 on the 18-19th and 22nd-24th. Thermometers had generally climbed into the teens after the first cold wave when the second blast out of the northwest hit about noon of the 22nd. January 23 became known as the bitterest day of the century with a greater windchill than on Cold Friday in 1810 or any other day in New England history since the introduction of the thermometer.

The readings taken at Craftsbury near the Canadian border were representative of conditions in the north country. The standard Smithsonian thermometer read −34°, −23°, and −27°, for a mean of −28°. Burlington in the Champlain Valley was only slightly warmer, with a low of −30° and a mean of −23° for the day. Both figures were the lowest in the thirty years of observations on the University of Vermont campus. The Dartmouth thermometer at Hanover, New Hampshire, on the 23rd registered the following: −26°, −18°, and −35°, for a mean of −23.3°, also the coldest recorded day there.

On the morning of January 24, Vermont thermometers generally reached all-time nadirs. Charles Marsh of Woodstock, a leading citizen of his town who served as the Smithsonian observer, commented in his notes to Washington: "On the morning of the 24th Jan. all mercurial thermometers in the vicinity were frozen. The thermometer furnished me by the Institution had not been received and there was no spirit thermometer nearer than Brownsville (15 miles distant). That I believe indicated 43 below."

196

Another communication to the *American Journal of Science* by Professor Young contained an interesting commentary:

> The mercury did not quite congeal in our thermometers on the morning of the 24th of January, but would doubtless have done so at the Hanover bridge, in the valley, 140 feet lower than our village, had not the owner of the instrument, on perceiving the rapid descent of the mercury on the evening of the 23d, taken it in, to prevent its freezing and bursting. He was not aware that mercury, unlike water, contracts, on congealing, instead of expanding. In low grounds surrounded by precipitous hills the temperature, when the air is still, is generally some degrees lower than in more elevated localities. Thus at the White River Junction (which is about 187 lower than our village), the mercury is reported to have frozen.

A correspondent of the *Coos Republican* at Lancaster, New Hampshire, wrote: "Jan. 24—The coldest day I ever witnessed. Thermometer at sunrise at store 55 degrees below; at 8:00 A.M. 45 degrees below; at 9:00 A.M. 38; at 10 A.M. 35 & at 11 A.M. 22 below; a thick mist or fog prevailed till 10 A.M." The editor of the newspaper commented: "The thermometers all froze up and almost everything else."

The *Daily Free Press* at Burlington commented in an article entitled: "The Cold Friday of 1857":

> Yesterday was the coldest day on record at Burlington. The thermometer, as noted from the instrument formerly used by Prof. Thompson in his observations, ranged as follows: At sunrise $-28°$; at 9 A.M. $-28°$; at 1 P.M. $-18°$; at 3 P.M. $-16°$; and at 9 P.M. $-27°$; giving a mean temperature for the day of twenty three and 4/10 degrees below zero. [The coldest day in the preceding nineteen years was February 6, 1855, the mean of which was $-19°$.]
>
> A strong wind from the North-West added to the severity of the temperature and made it almost insufferable.

Another Record Cold Day in January 1859

The record cold marks set in northern New England in 1857 were either equaled or exceeded two years later. Although it was a cold and snowy period, the first month in 1859 can in no way compare with the great combination of wintry elements that makes January 1857 the coldest month of the century.

Burlington's mean temperature of $-27°$ on January 10, 1859, can find no equal in the records maintained there since the 1830s. Professor Petty's thermometer on the campus of the university read: $-31.5°$ at 7:00 A.M., $-26°$ at 2:00 P.M., and $-25°$ at 9:00 P.M., for a mean of $-27°$, or four degrees lower than two years before. The *Free Press* published the figure as $-32°$ and that seemed representative of the area.

The intensity of the reservoir of cold air overlying northern Vermont and New Hampshire was apparent from the thermometer readings on a Smithsonian thermometer at Craftsbury. The mercury went below zero on the morning of the 8th and did not again attain that mark until the morning of the 12th. The coldest day occurred on the 10th with readings of $-27°$, $-20°$, and $-31°$; the absolute minimum of $-35°$ came on the morning of the 11th.

On lower ground at St. Johnsbury, elevation 510 feet, in a river valley, a Smithsonian instrument read by Franklin Fairbanks showed two remarkably cold days: January 10 $-26°$, $-19°$, and $-30°$; January 11 $-40°$, $-16°$, and $-22°$.

To the south in central Vermont at Woodstock, Charles Marsh rose early each morning to see what his thermometer was up or down to:

> The cold of the morning of January 11 has rarely, if ever, been experienced in this immediate vicinity. The mercury stood at $-39°$ from 5:00 A.M. (when I first saw it) to 7:30 A.M. Then it began rising slowly, the sky having become partly overcast with stratus clouds. A spirit thermometer near the mercurial standard at the same time indicated a temperature of $-45°$, but the mercury exposed in an open glass near the instrument did not congeal. The mean temperature of the day was $-25°$ which is the lowest mean recorded in this vicinity within my knowledge.

Conditions at Burlington were described by the *Free Press*:

> Unexampled Cold. – This morning from 6 to 8 o'clock the temperature (marked by the thermometer of the late Prof. Thompson of this place, the same instrument used by him for the last nineteen years of his life) was $-32°$, or 4 degrees colder than the morning of the famous cold Friday, Jan. 23, 1857, which was colder than had been experienced in this place for a great many years before. We are sure from our own observations for the last thirty-three years, that not so great a cold has been felt in Burlington within that time, by at least three or four degrees. We think the temperature has not been above $-23°$ or $-24°$ to-day.
>
> *The Daily Free Press* (Burlington), Jan. 10, 1859.

The Spectacular Temperature Surge in February 1861

February 6, 1861, provided northern Vermont with an ideal midwinter day – the type one might expect during maple sugaring time in March or early April. The mercury at both Burlington and Craftsbury climbed just above the freezing mark under the urging of a southerly wind flow.

But at that very time not far to the northwest in Quebec a cold front was making rapid progress in an impetuous rush from the arctic region to the tropics.

In the Champlain Valley, Burlington residents went to bed on the evening of the 6th with the thermometer at an even 32°. Next morning it was down to 5° and continued its descent all day: down to −6° at 2:00 P.M. and to −19° at 9:00 P.M. The reading on the morning of the 8th was −32°, to start a very bitter day; the three observations were −31°, −19°, and −19°, for an average of −23°.

For the elevated plateau and mountains of northern Vermont, the readings at Craftsbury were representative. After a cloudy day on the 7th, the thermometer dropped from 7° to −22° by evening, and next morning stood at −36°. The chill readings for the day were −36°, −22°, and −20°, for an average of −26°.

It was even colder in the upper Connecticut River Valley at Lunenburg. Dr. Hiram Cutting observed his mercury frozen on the morning of the 8th; his observations that day were −40°, −22°, and −30°, for an average of −30.6°.

Farther south at Hanover, New Hampshire, the gyrations of the mercury were of the same order, but more concentrated as to time. From 37° at 1:00 P.M. on the 7th, the plunge went to −32° at 7:00 A.M. on the 8th—a fall of 69 degrees in eighteen hours.

Some particulars of the cold wave were carried in the *Free Press*:

> The Weather. — Yesterday was one of the most uncomfortable days we have had in a long time. The storm, which it will be seen from our telegraphic dispatches was wide spread, raged here with great fury. Few persons were in the streets during the day, and fewer still in the evening. At 10 P.M. yesterday the thermometer indicated eighteen below zero, the wind blowing a gale. This morning at 7 o'clock, the thermometer stood at twenty-nine below zero. The wind has pretty much ceased, and the sun shines, but not with much effect upon the general temperature. The mercury stood at − 18 one o'clock.
>
> The coldest temperature of last winter was −13° on the 1st of February. The coldest temperature of 1859 was −32° on the 10th of January, the coldest ever noted in this place. In 1857, Jan. 23d, the thermometer stood at −28°, one degree higher than this morning.
>
> At White River Junction the mercury showed −30° this morning, Having fallen 41 degrees in five hours! At Rouse's Point it was −36°.
>
> *The Daily Free Press* (Burlington), Feb. 8, 1861.

The Cold Term of January 1866

An anticyclone of great physical dimensions and geographical extent made meteorological history during the first days of January 1866. New records for high barometric pressure were established from Iowa to New England; the high reading of 31.38 inches at Toronto, Ontario, on January 8, 1866, has never been equaled or even approached since. Severe cold of unprecedented duration accompanied the movement of the high-pressure area from central Canada into northern New England.

Barometric pressure began to rise as cold air entered the Northeast on January 4. Winds, however, were light and continued so. Zero readings reached Craftsbury by 9:00 P.M., the mercury having dropped from an afternoon maximum of 26°. From the evening of the 4th until late morning on the 9th, the temperature stayed below zero at Craftsbury; the highest during this period at an observation time was −8°. For over 100 hours the extreme cold endured, and at some cold pockets such as Lunenburg, it continued past the 9th, making 125 hours of below-zero readings.

The Daily Free Press of Burlington on January 10 took notice of the unusual cold term:

> The Great Cold Snap of 1866—Prof. Petty confirms our impression that the recent cold spell was entirely without precedent. We have had colder days, but never four such together. The mean temperature of last Friday [the 5th] was, discounting fractions, −11°, of Saturday, −14°, of Sunday, −22°, and of Monday, −17°. The mean temperature of the four days was sixteen below zero. There is nothing like that on the record, and we trust will not be again very soon.
>
> The thermometer this morning (the 10th) at sunrise stood at zero.

Cold Christmas of 1872

The following extracts from the records of three observers in Vermont indicate the extremity of the cold on Christmas Day 1872 and on the days following. The holiday has never been so severe until the recent "Cold Christmas of 1980" when similar conditions prevailed:

> *Randolph:* Dec. 25. Ther. 40 below at 7 A.M. A neighbor looked at his thermometer at midnight and it was 38 below. Mercury froze. Thinks a spirit thermometer would have indicated 44 or 45 at 7 A.M. at the usual rate of falling from midnight.
>
> *Lunenburg:* Dec. 25, −43°, mean for day −26.2°. Coldest ever known.
>
> *Woodstock:* Dec. 25—Spirit thermometer indicates −41.5°
>
> *Stratford, N.H.* (on the Connecticut River opposite

Maidstone): Christmas Eve. Temp. at 10 P.M. 30° below. Morning temp. at Lancaster House 46° below, at E. Savage's 52° below, at Mr. Ray's 55° below. Coldest morning since 1857. *History of Coos County*.

Smithsonian reports in National Archives.

The Cold Decade of the 1880s

Beginning with December 1880, a remarkable string of cold winters marked the progress of the decade until the winter of 1887-88. Based on the Burlington mean temperatures, all winters with the exception of 1881-82 ran either well below normal or produced a single severely cold month. Only one other of the eight winters under consideration averaged above normal, that of 1883-84, and only by a small margin.

The coldest winters ran from January to March 1885 and from January to March 1888. The coldest single month came in January 1888 when temperatures were 8.7 degrees below normal, and February 1885 was a close second with 8.3 degrees below normal. The lowest temperatures each winter were not extreme; a minimum of −23° was registered at Burlington in 1886 and 1888.

January 1888 also won honors as the coldest month in the Hanover records that run from 1835 to the present; a mean of 6.8° was 10.2 degrees below normal.

Burlington: Departure from Normal of Mean Temperature:
Winters 1880-81 to 1887-88

	Dec.	Jan.	Feb.	Mar.
1880-81	−2.1°	−4.6°	1.9°	9.8°
1881-82	11.5	0.6	6.2	3.5
1882-83	1.9	−3.9	2.2	−6.5
1883-84	−0.2	−5.7	4.5	−9.0
1884-85	3.6	1.6	−8.3	−9.6
1885-86	2.6	−1.1	−2.3	−0.3
1886-87	−3.0	−1.1	−1.8	−2.5
1887-88	0.2	−8.7	−1.9	−2.7

The Cold Turn of the Year: 1917-1918

The most extended term of intense cold in modern (1885 to the present) records sent Vermont into a deep freeze from Christmas afternoon of 1917 until Twelfth Night on January 7, 1918. Temperatures throughout the state were continuously well below freezing, and below 10° from December 29 through January 5 without respite. On the last three days of December, the mercury did not rise above zero during the daylight hours on any official thermometer throughout Vermont. Unfortunately, this was the only winter during America's participation in World War I, and the suffering from cold hit harder with scarcities of fuel.

Burlington in the northwest is usually the first large urban area in New England to feel the effect of a cold wave borne on the wings of northwest gales from the tundras and frozen lakes of Canada. The Queen city thermometer went below zero at 2:00 P.M. on December 28 and did not rise above that symbolic mark until noon on January 1, a period of ninety-four hours.

As to continuous freezing, Burlington's thermometer remained below 32° from December 26 through January 11 when it rose above for a few hours to a high of 39°; then another cold blast from Canada reestablished the freezing regime, keeping the mercury below 32° for a full month until February 11. The three winter months of 1917-18 averaged 12.1°, for the coldest full winter in the Burlington records from 1883 to the present.

December 30, 1917, appears to have been the coldest day Vermonters have endured this century, joining the select circle with January 29, 1780 and January 24, 1857, as the coldest days of their centuries. Several stations reported frozen mercury: St. Johnsbury, Bloomfield, and Hyde Park at −43°, Cavendish at −42°, Northfield at −41°, and Chelsea at −40°.

Burlington had daily mean temperatures of −21.5° on the 29th, −19.0° on the 30th, and −13.5° on the 31st. All minimums were zero or below through January 4. For the seven days, December 29, 1917, to January 4, 1918, the mean of all readings averaged −10.2°, the coldest week so far of the twentieth century.

Cold Waves of the Winter of 1933-34

The skiing industry in Vermont received its first major impetus with the construction of the first rope ski tow near Woodstock in 1934—the year that Vermont enjoyed one of its most distinguished winters for cold and snow.

The first cold wave of the season arrived on December 8 when the Burlington thermometer fell from an afternoon maximum of 35° to a minimum of −2° the next morning. During the following week the mercury dropped to 0° or lower each morning. The cold then relaxed for a few days until the arrival of one of the severest arctic outbreaks of record. Burlington was enjoying a balmy Christmas Day with the thermometer at 43° at noon when the wind shifted into the northwest with the arrival of a sharp cold front. By midnight the mercury had slumped to 7°, and next morning found it at −3°. The minimum continued below zero through the 29th with daily readings of: −12°, −16°, −29°, and −29°. The last day of the series produced one of the bitterest days in many a year: maximum −16° and minimum −29°, for a mean of −22.5°.

The all-time record minimum for Vermont, and also all New England (during the Weather Bureau period of coordinated records) occurred at Bloomfield on the upper Connecticut River where the thermometer dropped to an even −50° on the morning of the 30th. Other low readings were: −44° at Pittsburg, New Hampshire, −42° at St. Johnsbury, −40° at Newport. Deep snow cover existed throughout the state from December falls that averaged 22.5 inches and reached a maximum of 39.5 inches at Enosburg Falls in the northwest. The mean temperature in December for the state averaged 16.9°, or 5.5 degrees below normal.

Cold Friday III: February 9, 1934

January 1934 produced a near-normal month with a temperature mean only 0.4 of a degree below normal, and an average snowfall of 17.4 inches. After a cold first four days, a mild period ensued until after mid-month. Cold air invaded the state on the 17th-18th and dropped the mercury below zero for five days. Another cold wave arrived on the 28th, the start of a record-breaking cold term. Thereafter, until March 2, the thermometer never rose above 34° at Burlington, and on twenty-two of the thirty-two days the minimum fell below zero.

February 1934 went into the record books as the coldest month of any name in the Weather Bureau's experience to that time, with an average for the state as a whole of 5.8°, or 12 degrees below the current long-term normal. The coldest day came on the 9th in Vermont and throughout the Northeast where most urban cities along the seaboard set their all-time minimums.

The seventeen reporting stations in Vermont had an average minimum of −31.95° on the 9th. The lowest reading was −41° at both Bloomfield and East Barnet, and the highest minimum, −22° at Brattleboro. Burlington endured two bitter days: on the 8th, maximum −9°, minimum −25°; on the 9th, maximum −9°, minimum −28°.

The Spectacular Temperature Drop of 1943

On the afternoon of February 14, 1943, the thermometer at Bennington in the southwest corner of the state reached 45°. From that peak it fell steadily all night and the next day to reach a record −32° on the morning of the 16th. Two more below-zero mornings concluded Vermont's most spectacular cold outbreak of the twentieth century. Only February 1861 can match this for swiftness of temperature drop and the thermal depths reached.

Temperature descents of extreme proportions were reported throughout the state. East Barnet in Caledonia County registered − 46° on the morning of the 16th to lead all stations in Vermont and New England. But Bloomfield in nearby Essex County was only a degree behind at − 45°, followed by − 43° at St. Johnsbury and Woodstock, and − 40° at Cavendish.

The alcohol in the Weather Bureau thermometer at Newport climbed to only a maximum of − 19° on the 16th, for the coldest day of the century with a mean of − 24.5° on the shores of Lake Memphremagog astride the Canadian border.

The cold term continued until the 19th when most Vermont thermometers went above freezing in the afternoon. The recovery at East Barnet was almost as spectacular as the descent, rising from − 46° on the 16th to 55° on the 23rd − a climb of 101 degrees! On that day Enosburg Falls hit a high of 60°, up from a nadir of − 39° a week earlier.

Despite the extreme cold wave, the month of February 1943 averaged 1.7 degrees above normal, in much the same fashion as the elements behaved in January 1810 when Cold Friday occurred in an otherwise mild winter season.

Cold January of 1957

January 1957 stands as one of the notable cold months of the century in Vermont and throughout New England. The mean temperature for the state was 10.4°, or a departure from a normal of 6.3 degrees for the thirteen reporting stations. Most of the month ran rather true to form with the only warm period coming during a January thaw at the traditional time, from the 21st to 23rd, when a maximum of 51° occurred at Burlington. At this time Bennington registered the highest readings in the state with 50°, 60°, and 60° on successive days.

The hard core of the cold prevailed from the 14th to 21st; West Burke in the Northeast Kingdom dropped to daily lows of − 38°, − 40°, − 40°, − 22°, − 30°, − 30°, − 15°, and − 12° during this period. The minimum fell below zero every night from the 1st to 21st, except on the 8th, 10th, and 13th, and on these days the nighttime minimums ranged between 0° and 5°. After two-and-a-half days of thawy weather from the 22nd to 24th, it turned cold again at West Burke, with four more below-zero nights at the conclusion of the month. The mean temperature for January of 3.2° was of arctic proportions.

It's a sad day when pipes freeze in the house. It's equally disastrous when road culverts freeze solid, for highway crews know that without open culverts the spring runoff can wash away the finest stretch of road. Here a state highway team uses steam to unplug a frozen culvert.

Burlington had a record day on the 15th when the mercury remained below zero all day, ranging from a minimum of −30° to a maximum of −6°, for a mean temperature of −17°. The minimum equaled the twentieth-century record for the Queen City.

Absolute minimums for the month were reminiscent of the cold days of the mid-nineteenth century: Lemington −42°, Enosburg Falls −41°, West Burke −40°, and Mt. Mansfield −39°.

Cold January of 1961

As January 1961 developed, it became the first month to average below 10° at Burlington since the famous month of February 1934 when the all-time monthly low of 5.6° was established. The January 1961 mean at the Queen City ran 8.7 degrees below normal, and followed a December averaging 3.1 degrees below normal, making a very cold start of winter.

All over the state, January 1961 averaged 10°, amounting to a depar-

205

ture of 6.7 degrees from the normal. West Burke dropped to zero or below on all but four nights in January, and on every night from January 10 to February 9, a full thirty-one days of unbroken zero nights. The lowest readings of the winter were reported at Mt. Mansfield, − 33°, and at West Burke and Bloomfield, − 32°.

After several cold days at the start, February turned milder to average 0.8 of a degree above normal, ending the possibility of an all-time cold winter to surpass 1917-18.

Cold January and February of 1968

The winter of 1967-68 was the first of a series of winters that gave Vermont a real arctic experience for four seasons in a row. The month of January 1968 averaged 6.6 degrees below normal and February 4 degrees below normal for the state. At Burlington, the two months combined made up the coldest sixty consecutive days since the current records began in 1883.

The core of the cold occurred between January 7 and 12 when Queen City residents endured continuously below-zero temperatures for 115 hours, with successive morning minimums of − 9°, − 21°, − 24°, − 22°, − 27°, and − 23°. The maximum during this frigid sequence rose only to − 9°. At the same time, West Burke had below-zero nights from the 6th to 19th, with a concentration of extreme low readings from the 9th to 13th of: − 34°, − 32°, − 32°, − 34°, and − 34°. The average minimum for the entire month at West Burke was − 10.7°.

In February 1968, except for an aberration on the 4th and 5th, all minimum readings at West Burke were less than 10°. On twenty-one of the twenty-nine nights readings dropped below zero. The minimums were not as extreme as in January, the lowest being a mere − 25°. The coldness inhibited snowfall; the average statewide for January was 17.7 inches and for February only 9.9 inches.

The Continental Month: January 1970

The coldest month of record at Burlington was also one of the least snowy, conditions more likely to be experienced on the tundra of central Canada than in the Green Mountain region. The mean temperature of 3.6° figured two whole degrees colder than the past record coldest month in February 1934, and it ran colder than the previous coldest January in 1920 by 2.8 degrees. The total precipitation of only 0.65 of an inch comprised the fourth driest January of the century. Only 11 inches of snow

Ice skating, next to snowshoeing, was probably the most popular winter sport in Vermont before the days of skiing. Brooks, ponds, lakes, rinks, and even flooded fields offered skating opportunities for all the members of the family. Ice skating was a graceful sport which people of all ages could enjoy.

fell during January in a season which otherwise produced a total of 104.6 inches. (And January normally makes the greatest contribution to the season's snowfall total).

Burlington's minimum thermometer sank below zero on December 22 and continued to drop below that mark on each night until January 9. On the 10th, the cold eased momentarily with the thermometer ranging from a low of 6° and to a high of 12°. Then the minimum again went into the nether region from the 11th to 16th. Another brief respite on the 17th with readings of 26° maximum and 16° minimum was followed by another series of below-zero nights through the 24th.

A record cold month was well in sight when a belated January thaw arrived on the 28th. The thermometer soared to maximums of 35° and 41° on the 28th and 29th. But the timely arrival of a cold front brought thermometers down to minimums of −3° and −5° on the final two days to assure the establishment of a new record.

The mean minimum for January at Burlington was a chill −7.5°. When combined with a mean maximum of 14.6°, the overall mean of 3.6° was 12.6 degrees below normal, and easily the coldest month in modern records at Burlington.

Despite the low-average coldness, the absolute minimums during the month were of modest proportions. Northfield and West Burke registered minimums of −28° for the lowest in the state, and Bellows Falls and Readsboro in the southeast reported −15°. All other stations ranged between these.

The Cold Winters of 1975-76 to 1978-79

As the winters from 1968-69 to 1971-72 were called "the snowy winters," those from 1975-76 to 1978-79 may be designated "the cold winters," since each season was distinguished by a severely cold calendar month. The mean temperatures at Burlington were: 11.1° in January 1976, 11.1° in January 1977, 9.5° in February 1978, and 7.5° in February 1979. Each January just missed the first ten ratings for the coldest first months, their standing being 11th and 12th. But each February was the 2nd and 3rd coldest in the Burlington records dating back to the winter of 1883-84.

The lowest temperatures statewide each winter were:

> January 1976, Enosburg Falls −34° on 24th
> January 1977, −35° Mt. Mansfield on 18th, −30° at Enosburg Falls and West Burke on 18th.
> February 1978, −32° Enosburg Falls on Feb. 4th.
> January 1979, −32° Enosburg Falls on 19th
> February 1979, −38° Enosburg Falls, −34° Chelsea, −32° St. Johnsbury on 12th, −37° West Burke on 12th and 13th

The Cold Term in February 1979

The first eighteen days of February 1979 appear to have been the coldest period of such length ever experienced in Vermont. Burlington's highest reading over these days was 16°, and the minimum dropped below zero on sixteen of the eighteen days. Four times the minimum sank to −25° or lower. The all-time modern minimum record for Burlington, −30° in 1957, was equaled on February 12. Throughout the region, minimums ranged down to −38° at Enosburg Falls in the northwest and to −40° at both Lancaster and Colebrook on the New Hampshire side of the Connecticut River. Dorset in the southwest went down to −30°, and in the southeast, Ball Mountain to −25° and Cavendish to −27°.

West Burke had the lowest average: maximum 17.3° and minimum −10.4°, or a mean of 3.5°−this despite a week-long thaw at the end of the month. The West Burke thermometer was continuously below freezing from the 1st to 20th. A most impressive string of low readings was put together from the 10th through the 21st: −33°, −33°, −37°, −37°, −28°, −30°, −30°, −21°, −33°, −33°, −18°, and −14°.

On a statewide basis, February 1979 with a mean of 9.0° ranked as the second coldest February behind 1934 with 5.8°.

Cold Christmas of 1980

Residents of Burlington awoke on Christmas morning in 1980 to find the official thermometer at −5°, down from a maximum of 31° on yesterday afternoon. The mercury continued to plummet during the daylight hours of the holiday to reach −25° by midnight. Next morning it stood one degree lower at −26°. With the passing of the core of the cold air overhead and a bright sun shining down, a rebound to a mild 14° took place in the afternoon of the 26th. This proved not only the coldest Christmas Day of modern times, but also was one of the bitterest days ever experienced with a very low windchill. The minimum also set a record for the coldest ever reported so early in the season.

Elsewhere in the state, Mt. Mansfield hit −38° on the 25th and −31° on the 26th, West Burke in the northeast dropped to −35° on the morning of the 26th. The cold wave covered all the Green Mountain region; Vernon in the southeast reported −20° on the 26th, and the mercury remained below zero all that day.

Cold Januaries in 1981 and 1982

The string of single exceptionally cold months marking the winters since 1975-76, with the exception of 1979-80, continued into the new decade. January 1981 averaged 8.0° statewide for the fifth coldest January in modern records (1895-1984), and January 1982 with 7.7° was the fourth coldest January. The only colder months than these two in the 90 winters were: February 1934 (5.8°), January 1918 (6.0°), January 1970 (6.5°) and January 1920 (6.8°).

In January 1981, the coldest day was the 4th with the following minimums: Enosburg Falls −41°, West Burke −38°, Northfield and Mt. Mansfield −35°, Montpelier and Morrisville −34° and Newport and St. Johnsbury −31°.

In January 1982, despite the lower average, the extremes were less: Enosburg Falls −33°, West Burke −31°, Chelsea −29°, and Cavendish −28°.

Brrrr! Enough of cold statistics.

Frozen Mercury in Vermont—

Extreme Minimum Temperatures Since 1900

1904, Jan. 19: −44° Enosburg Falls
1914, Jan. 14: −44° Bloomfield
1917, Dec. 30: −43° Bloomfield, Hyde Park, St. Johnsbury
1918, Jan. 28: −40° Enosburg Falls
1920, Jan. 31: −41° Enosburg Falls
1920, Feb. 1: −39° Enosburg Falls, Garfield
              −38° Bloomfield and St. Johnsbury
1925, Jan. 28: −42° Enosburg Falls
1933, Dec. 30: −50° Bloomfield   −40° Newport
              −43° East Barnet
              −42° St. Johnsbury
1934, Jan. 31: −44° Enosburg Falls
1934, Feb. 9: −41° East Barnet, Bloomfield
              −39° Enosburg Falls
1943, Feb. 16: −46° East Barnet
              −45° Bloomfield
              −43° Woodstock, St. Johnsbury
              −40° Cavendish
              −39° White River Junction, Enosburg Falls, Northfield, Somerset

1957, Jan. 14: −42° Lemington
            15: −41° Enosburg Falls
                −39° Mt. Mansfield
                −38° Bloomfield, Newport, Cavendish, Woodstock
            16: −40° West Burke
1962, Feb.   2: −41° West Burke
                −40° Bloomfield
             3: −41° West Burke
1963, Feb.   2: −40° Enosburg Falls
1965, Jan. 15: −39° Mt. Mansfield
1979, Feb. 12: −38° Enosburg Falls
1980, Dec. 25: −38° Mt. Mansfield
1981, Jan.   4: −41° Enosburg Falls
                −38° West Burke

Dates of Closing of Lake Champlain by Ice

1816 .... February 9	1840 .... January 25	1864 .... February 17
1817 .... January 29	1841 .... February 18	1865 .... January 17
1818 .... February 2	1842 .... Not closed	1866 .... January 30
1819 .... March 4	1843 .... February 16	1867 .... January 20
	1844 .... January 25	1868 .... January 7
1820 .... February 3	1845 .... February 3	1869 .... January 19
1821 .... January 15	1846 .... February 10	
1822 .... January 24	1847 .... February 15	1870 .... February 25
1823 .... February 7	1848 .... February 13	1871 .... January 24
1824 .... January 22	1849 .... February 7	1872 .... January 8
1825 .... February 9		1873 .... January 29
1826 .... February 1	1850 .... Not closed	1874 .... February 1
1827 .... January 21	1851 .... February 1	1875 .... January 16
1828 .... Not closed	1852 .... January 18	1876 .... February 2
1829 .... January 31	1853 .... January 28	1877 .... January 15
	1854 .... January 28	1878 .... January 29
1830 .... January ?	1855 .... February 4	1879 .... January 29
1831 .... January ?	1856 .... January 22	
1832 .... February 6	1857 .... January 15	1880 .... February 2
1833 .... February 2	1858 .... February 12	1881 .... January 16
1834 .... February 13	1859 .... January 11	1882 .... February 25
1835 .... January 10		1883 .... January 26
1836 .... January 27	1860 .... February 2	1884 .... January 8
1837 .... January 15	1861 .... January 23	1885 .... January 29
1838 .... February 2	1862 .... February 5	1886 .... January 19
1839 .... January 25	1863 .... February 4	1887 .... January 8

211

1888.........January 22	1920..........January 18	1953 .... Not closed
1889 .... February 7	1921........February 26	1954 .... Not closed
	1922..........January 24	1955 .... February 9
1890........February 21	1923..........January 28	1956 .... Not closed
1891..........January 27	1924........February 19	1957 .... Not closed
1892........February 14	1925..........January 20	1958........February 12
1893..........January 15	1926 .... February 9	1959 .... February 9
1894 .... March 26	1927.............March 4	
1895 .... February 8	1928 .... March 11	1960 .... Not closed
1896........February 17	1929 .... February 6	1961..........January 27
1897..........January 31		1962........February 16
1898..........January 30	1930..........January 30	1963 .... February 8
1899 .... February 1	1931........February 11	1964 .... Not closed
	1932.........Not closed	1965 .... Not closed
1900........February 28	1933.........Not closed	1966 .... February 7
1901 .... February 1	1934..........January 30	1967........February 13
1902..........January 30	1935..........January 30	1968........February 16
1903..........January 24	1936..........January 26	1969.............March 2
1904..........January 19	1937 .... March 12	
1905..........January 22	1938.............March 4	1970..........January 21
1906 .... February 7	1939..........January 27	1971 .... February 2
1907..........January 25		1972........February 10
1908 .... February 3	1940..........January 26	1973........February 20
1909 .... February 9	1941 .... February 1	1974 .... Not closed
	1942 .... February 5	1975......... Not closed
1910........February 11	1943........February 18	1976 .... Not closed
1911..........January 18	1944 .... February 9	1977..........January 16
1912..........January 26	1945..........January 25	1978 .... February 5
1913.......February 10	1946........February 21	1979........February 11
1914........February 11	1947........February 23	
1915........February 20	1948..........January 26	1980 .... Not closed
1916........February 20	1949 .... Not closed	1981..........January 14
1917..........January 27		1982..........January 28
1918..........January 24	1950........February 26	1983 ....Not closed
1919.........Not closed	1951........February 10	1984 .... Not closed
	1952 .... Not closed	1985........Not closed
		1986.......February 10
		1987.......February 18
		1988........Not closed
		1989.......February 22
		1990........Not closed
		1991........Not closed
		1992........Not closed
		1993........February 7
		1994........January 23

Source: National Weather Service Office, Airport, Burlington.

HEAT WAVES

Most people would not associate extreme heat with Vermont, but the thermometer does mount into the 90s annually and on a few occasions has topped the 100° mark. Since 1901, the century figure has been attained in fifteen of the eighty-one summers. A closer inspection reveals that only one instance has occurred in the past twenty-five years: 100° at Cornwall on August 2, 1975, the famous "Hot Saturday" in all New England. There were readings of 99° in both 1977 and 1978.

The all-time maximum registered on an official National Weather Service thermometer was 105° at Vernon in the southeast corner of the state on July 4, 1911, during one of New England's most extended heat waves that covered the first twelve days of that month. The second highest reading, 104°, occurred at Cornwall in the Champlain lowlands on August 21, 1916. Atop Mt. Mansfield, the highest reading in the past decade has been 81° on July 20, 1977 (record for the hot summer of 1975 is missing).

The Long Heat Wave in July 1911

Heat records that stood for many years were established in Vermont, New Hampshire, Maine, and Massachusetts during the first ten days of July 1911. The 105° reading at Vernon on July 4, 1911, and the 106° reading at Nashua, New Hampshire, on the same day are current records for extreme heat in the two bordering states.

The intensity and duration of the heat wave was well attested by two Vermont stations:

1911, July	1	2	3	4	5	6	7	8	9	10	11	12
Burlington 86°	94	100	98	98	91	75	88	95	96	96	86	
Northfield 84°	89	98	96	97	95	78	86	92	95	91	88	

The Heat of August 1944

Burlington's longest string of consecutive days with maximum readings of 90° or higher extended from August 10 to 17, 1944. The 101° on the 11th represents the highest ever attained at Burlington in records which date back to the 1830s. Only 0.04 inch of rain fell during this period until a thunderstorm deluge of 1.77 inches on the last day of August put an end to the combined drought and heat wave.

Five consecutive days with 90° or more at the opening of the month provided a good start in compiling a total of thirteen such days during the month, another still standing record.

The August mean of 72.1° was 4.2 degrees above the normal then in use, and 4.8° above the 1940-70 normal. July 1944 had averaged 72.1° also, or 1.8 degrees above the contemporary normal, and contributed three more 90° days. These figures made a very hot high summer and would have rated among the five hottest full summers if June had not been slightly below normal. Somerset, high in the mountains of southern Vermont at 2080 feet altitude, is included to show the effect of elevation on heat waves.

1944, August	9	10	11	12	13	14	15	16	17	18
Burlington	89°	95°	101°	91°	95°	96°	97°	96°	93°	78°
Bennington	85	89	96	95	97	85	92	97	90	82
Bloomfield	86	88	95	93	89	96	95	95	93	89
Somerset	79	84	88	87	88	78	87	88	87	67

1949—The Hottest Full Summer

Thanks to a record warm June, the summer of 1949 rated the hottest in Vermont state records. The three months, June to August, produced the following plus departures from normal: June 6°, July 3.2°, and August 2.3°, for an average departure of plus 3.8°.

In no other summer did the temperature at Burlington average above 70° for each of the three summer months. Extreme hot days with the maximum 90° or more numbered twenty-five, with June having six, July eleven, and August eight.

214

The highest temperature throughout the state during the summer reached 99° at Cornwall on August 10.

The summer of 1949 had the greatest number of cooling degree days, days which are based on the number of degrees the mean daily temperatures exceed 65°. Considering records from 1931 to the present, the total in 1949 was 478 cooling degree days, compared with a normal for a full summer of 294.5. In only six years has there been a total of 400 or more.

"Hot Saturday" in 1975

The outstanding heat wave of the 1970s occurred in the middle year of 1975. It featured "Hot Saturday," when new maximum records for all-time were established in Massachusetts and Maine, and closely approached in New Hampshire and Vermont.

On the last day of July 1975, temperatures in the Champlain Valley rose to 90° or above, with the highest of 95° at Cornwall, as usual. Next day all stations in the state except at high elevations in the southeast reached 90° or more. Cornwall, again, topped the list with 99°.

Saturday, August 2, proved the hottest day throughout Vermont since July 1911. Cornwall reported an even 100°, and most stations reached 95° or more. Burlington hit 99°, and Hanover (N.H.) soared to 103° for its warmest ever.

The fourth day of the heat wave, Sunday, saw Vernon in the southeast at 99°, and only locations in the extreme north failed to top 90°. All stations in Vermont, except Mt. Mansfield, reached 94° or more on either the 2nd or 3rd.

For the month as a whole, July averaged 71.1° over the state, or only 0.4 degree below the record high for any month set in July 1955.

Hot Mid-July 1979

A warm spell occurred in mid-July 1979 when the Burlington thermometer registered a mean of 72.2° for the month, or 2.4 degrees above normal. Two periods of extreme heat with 90° or higher readings extended from the 12th to 16th and 19th to 25th. From the 8th through the 16th and 19th through 31st, all maximum readings at Burlington were 80° or more. And the high readings continued into August through the 5th. The highest readings in the state were 96° at Vernon and 95° at South Hero, showing that the heat was widespread.

1909, Aug.	8:	100°	St. Johnsbury
1911, July	3:	101°	St. Johnsbury
July	4:	100°	Woodstock
July	5:	102°	Cavendish
1913, July	1:	101°	Cavendish
1916, July	21:	104°	Woodstock
Aug.	21:	104°	Cornwall
1918, Aug.	7:	102°	Vernon
		100°	Cornwall
1919, June	4:	101°	St. Johnsbury
		100°	Cornwall

1920, Aug.	11:	102°	Vernon
1921, July	8:	101°	Cornwall
1926, July	22:	102°	Brattleboro
1933, Aug.	1:	100°	Cornwall
1937, Aug.	30:	100°	Chelsea
1944, Aug.	11:	101°	Burlington
Aug.	15:	100°	Wilder
1953, July	18:	100°	Cornwall
July	19:	102°	Bellows Falls
1956, June	15:	100°	Bellows Falls
1975, Aug.	2:	100°	Cornwall

216

FLOODS

The behavior of the atmosphere over Vermont can create no more devastating threat to its well-being than the condensation of water vapor floating in clouds aloft into a torrent of descending water droplets. After splattering the high slopes of the Green Mountains, these droplets coalesce into a fluid mass whose ever-growing volume seeks a channel down to the main waterways that drain the various portions of the state in four different directions.

When descending the narrow defiles of the upper slopes, the mass of water is given an irresistible force by gravity, causing extensive erosion of soil and ground cover. Fields of crops are covered and roadways washed away.

After several streams have poured their discharges into a single river valley, the rising water level inundates riparian land and property. The surging waters carry an assemblage of debris, composed of limbs of trees, fragments of buildings, spans of bridges, and any other loose object that is buoyant. Sometimes huge cakes of jagged ice augment the floating flotsam. Together these serve as battering rams to smash and destroy any obstacle along the way such as dams, bridges, and mills. Riverfront buildings are undermined and sometimes collapse to join the floating debris, adding to its destructive power. Railroads and highways are severed, and lines of communication and power cut. Whole communities become isolated.

Floods may occur in Vermont under a variety of meteorological situations. Usually a flood is preceded by an air flow from the moist sources of either the Gulf of Mexico or the Atlantic Ocean. Cyclonic systems raise moisture from the surface of the ocean and transport it in the form of water vapor many hundreds of miles until it condenses into clouds and precipitates rain drops or snow flakes. Transcontinental cyclonic storms moving along the Canadian border from Alberta do not have sufficient moisture to create flood conditions since they draw upon continental interior sources. They usually pass down the St. Lawrence Valley stormway, giving the Green Mountains only a brief period of showers or light precipitation as a cold front approaches and passes.

Storms originating in Texas or over the western Gulf of Mexico move northeast, and often prove good rain producers over the eastern United States. They bring north a flow of tropical air conditioned over the warm, moist surface of the Gulf of Mexico. Often storms from Texas and the Gulf area are accompanied by an active warm front that lifts the moist airstream over residual cool air overlying New England, causing constant condensation and continuous heavy precipitation. If a Gulf storm center passes south of Vermont over the Middle Atlantic States and along the shores of southern New England, the Green Mountain area can be assured of moderate to heavy rainfall.

Another potential flood producer is the coastal storm that originates along the shoreline or over the offshore waters of the southern Atlantic states and moves up the seaboard as a northeaster. If the path of the storm center passes inside or over Cape Cod and the Islands, southern Vermont, at least, is likely to receive a copious rainfall. Following similar tracks are the tropical storms and hurricanes in the summer and fall seasons.

Some of Vermont's greatest floods have resulted from tropical moisture evaporated from ocean surfaces a thousand miles away and brought to New England by tropical storms, either in their full hurricane stage or by the dissipating circulation of a hurricane. When Vermont lies to the west of a hurricane track moving over eastern New England, the possibility of a rainfall with flooding is present.

Local conditions also may give rise to a flood situation. If a moist, unstable air mass is present, thunderstorms may form and develop into a cloudburst, giving rise to a flash flood. When thunderstorms remain stationary over an area for one to two hours or move very slowly, very heavy amounts of rain may fall over a restricted area. Thunderstorms also can form on a stalled frontal system and continue to produce heavy showers for substantial periods.

Antecedent weather conditions are all-important in creating a flood potential. Cyclonic storms in winter usually do not carry more than an inch or two of precipitable water, and a single storm does not cause enough rainfall for a flood. But a series of two or three rain periods occurring within a week or ten days can saturate the soil enough to cause a heavy runoff when a hard rain arrives. The Vermont Flood of 1927, the March Flood of 1936, and the Hurricane Flood of 1938 were all preceded by periods of moderate to heavy rainfall. Also in late winter and

218

early spring, the water content of the snow cover remaining on the ground is an important factor. When a warm rain falls on deep snow, the amount of water released by the melting snow may be considerable and greatly augment the runoff from the rainstorm.

Floods may occur at any time of the year in Vermont. January and February sometimes feature floods when a midwinter thaw causes a sudden break-up of river ice with subsequent jamming and temporary flooding. March and April usually create "spring freshets," which result from a combination of gradual snowmelt and a period of warm rains during a cyclonic storm. Flash floods are the mark of the summer months when thunderstorms cause local deluges and quick rises on small streams. During the summer doldrums a stalled front across the Green Mountains has produced two to three days of heavy rainfall that comes in bouts of showers at intervals that flood all rivers. September is the month of a hurricane-spawned flood such as occurred during the Great September Gale in 1815 and the New England Hurricane in 1938. October and November have brought lesser tropical storms northward with their excessive loads of maritime moisture, as occurred in October 1869 and November 1927. December is the least likely month for a Vermont flood.

The First Recorded Flood: January 1770

The historian of New England's famous storms, Sidney Perley, included the big rainstorm on January 6-7, 1770, in his list of outstanding weather events, calling it "the greatest freshet perhaps that ever occurred in New England."* Though this is probably a bit of hyperbole, the rain pattern was extensive, covering most of the river valleys of the region: the Androscoggin and Kennebec in Maine, Merrimack in New Hampshire, and the Connecticut in western New England. All were in high flood on the following days.

In southeast New Hampshire, "as Great a Flood of Rain as has been Known in the Age of man swelled the fresh rivers so as to sweep away most of the Bridges over them hereabouts and in this Town in Particular (Rochester, N.H.) and many mills were Carried away or much Damaged," according to the town's historian.

The effect of the storm reached west into Vermont with similar impact, though local details are lacking. The compiler of the *Annals of Brattleboro* noted that the flood-prone little island in the Connecticut River opposite the village, always employed as a flood level indicator, was under water in 1770, presumably in this January flood.

In the lower Connecticut River Valley, the editor of the *Connecticut Courant* gave notice of conditions at Hartford: "In consequence of the

*Sidney Perley, *Historic Storms of New England*, Salem, Mass., 1891.

plentiful rain (which began on Saturday night the 6th inst. and continued till Monday morning), there has been such a sudden and extraordinary rise of the water of Connecticut River, as been rarely known, by which all the low land adjoining has been flooded."

The Independence Flood in October 1775

A vast rainstorm of historic importance spread over most of New England from October 18 to 21, 1775, when the opening phases of the military operations of the War of Independence were taking place. The invasion of Canada by two American forces was under way, and both experienced the severity of the storm.

Reports of the violence of the winds, the excessive rains, and the tremendous rise of the waters of small streams in a very short time came from locations in northern Maine to southern Connecticut. The historian of Keene in southwestern New Hampshire related: "This remarkable great rain caused a great flood," and a diary entry for October 21 by Matthew Patten at Bedford, also in the southern part of the Granite State, told of "an exceeding Rainy Day, great flood in the river."

No details of the extent of the inundations or damage for Vermont localities have been uncovered since the newspapers of the day were filled with military and political items. But down at Hartford, the *Connecticut Courant* referred to: "the most sudden and extraordinary freshet in Connecticut River ever known at this season of the year, by which much damage has been done to fields of corn in the meadows, and vast quantities of hay and flax entirely carried off." Such a great volume of water could only have originated in the upper reaches of the valley in Vermont and New Hampshire.

A Spring Freshet in 1783

The initial issue of the first newspaper to be published in Vermont carried the story of a freshet on the Castleton River, which flows into the Poultney River and eventually into the East Bay River near Whitehall, New York. *The Vermont Gazette, or Freemen's Depository*, published by Anthony Haswell and David Russell at Bennington, reported the event on June 5, 1783:

> By a gentlemen from the Northward we are informed that they have lately had extremely heavy rains in those parts, during the continuance of which, a large river which runs through Castleton, &c. and empties itself into the East-Bay, took a sudden turn from

its old channel and has worn a new one of an amazing depth. Many lofty pines (some supposed to be 100 feet in length) hang by the roots, but their tops seem to be a great way from the bottom. By this sudden change the falls of the Castleton are left entirely dry. Our informant further says, that there has been very heavy rain on the west road through Wells, Poultney, &c. to Castleton, which has entirely destroyed many bridges, and very much damaged others.

October Snowmelt Flood in 1783

Three storms, probably of tropical origin, swept northeast along the Atlantic seaboard during October 1783. Their combined effects caused a damaging flood in the Green Mountain region.

After lashing Charleston, South Carolina, on October 7-8, the first hurricane moved up the coast and battered New York and the ports of southern New England on the 8-9th. Interior New England received the heavy rain that usually falls to the west and north of the track of a tropical storm. On the 18-19th of October, an even larger tropical storm with hurricane-force blasts moved northeast along the same route. Its vast circulation, drawing on a reservoir of cold air over Canada, treated most of New England, except for the coastline, to a substantial early-season snowfall.

An early settler at Woodstock judged the depth at 12 inches there, with even greater amounts on the Green Mountain ridges to the west. These precipitation amounts laid the basis for a vast runoff when a third storm of heavy rain arrived on the 23rd. It was "the greatest flood ever known" in the opinion of the editor of the *Vermont Journal and Universal Advertiser*:

> On Wednesday night [Oct. 22], last week, we had a most heavy storm of rain, which, with the melting of the snow that had fell a few days before and was lying 6 or 8 inches deep in the woods and on the heights of land, raised the streams to such a degree as produced the greatest flood ever known since the settlement of this country. Several mills and mill-dams in this town and its vicinity were carried away: a number in Claremont were destroyed: and indeed no town in this part of the country, on either side of Connecticut River, has escaped losing several. The bridges likewise all over the country were so universally washed away, that there has been but little travelling since, especially before the streams lowered. The grist-mills are rebuilding and repairing with all possible expedition; and we are happy in observing, that it is the general opinion people will not suffer so much for want of bread as was at first feared.
>
> *Vermont Journal* (Windsor), October 30, 1783.

"The Great Freshet of October 1785"

> A considerable amount of rain fell in the month of September, 1785, and from time to time during October it continued to fall in unusual quantities until Thursday, the twentieth. The rain descended through that day, and in the evening the wind shifted from west-northwest to the opposite direction, blowing hard through the night. The wind continued to blow from the east-southeast for two days, and during this time the rain steadily fell in extraordinary quantities. The storm cleared up about ten o'clock on Saturday night (the twenty-second), nine inches of water having fallen during the three days. It fell principally in southeastern New Hampshire and the adjoining country, and was the heaviest fall of rain on record that has occurred in New England in so short a space of time. It caused a great freshet in the region that it covered and proved exceedingly distressful to the inhabitants.

With these words Sidney Perley in his *Historic Storms of New England* described what appears to have been the greatest all-New England flood of the early period and a worthy rival to that of March 1936.

Town histories of New Hampshire and Maine declare the flood of October 1785 was the greatest ever known. A tabulation of maximum discharges of the Merrimack River at Lowell, Massachusetts, shows that of October 24, 1785, exceeded all others of the eighteenth or nineteenth centuries.

The impact of the storm in Vermont appeared to have been greater in the northern part than in the southern. Colonel Thomas Johnson at Newbury on the Connecticut River left this entry: "Oct. 23. This day the great freshet was at the height which covered all the meadows— swept all the fences off." Daniel Thompson in his *History of Montpelier* devoted considerable space to the local impact of the flood of 1785:

> From incontestible indications, it appears that the water in that unprecedented rise of the Winooski (1785), rose some three or four feet higher than the highest parts of State Street. This would have submerged nearly every acre of the whole of the present site of Montpelier Village, to depths varying from one to a dozen feet, from the rise of the hills on one side to that of the corresponding one on the other side. Should such another flood occur what destruction of property must ensue?

Since Montpelier was not settled until two years after the flood of 1785, no property damage occurred. But as Thompson feared, a flood of similar proportions in November 1927 brought devastating consequences.

The "Sea of Blood" in October 1786

Red came the river down, and loud oft
The angry spirit of the waters shrieked.

John Andrew Graham

A storm and flood attended by unusual circumstances struck western New England on October 5-6, 1786. It was reported severe in the hills around Litchfield in western Connecticut where "an exceedingly heavy southwest wind and heavy rain" did damage on the 5th. The excessive downpours extended north along the crest of the Berkshires and Taconics.

A spectacular incident during the storm took place in Arlington, Vermont, southwest of Manchester, when a flow of rain water was dammed high on the side of a mountain slope whose soil was composed of red sandstone earth. On finding a channel of release, the now red-colored torrent rushed down the mountainside into Roaring Brook.

The only contemporary account of the event appeared in a letter from a resident of Berkshire County in Massachusetts to a New London newspaper. The article made the rounds of the country's press and appeared in the *Columbian Magazine* of Philadelphia, the country's first news periodical:

> On the 5th of October we had a most extraordinary storm of wind and rain, which raised the rivers to a height scarsely ever known before; the mills and bridges in many towns are almost all damaged or gone, and the destruction of hay and corn in the meadows is very considerable. On the 6th day, in the morning, there was a noise, something like an earthquake, heard in Manchester, state of Vermont, when on a sudden a flood rushed from the west mountain, and ran with such violence in a breadth of about 26 rods wide, that it was judged, where the mountain was as steep as the roof of a common house, the water ran near ten feet deep, throwing the timber into vast heaps, and washing out rocks of many tons weight; and carrying down large quantities of red paint, with which the mountain abounds, forming in the meadows and streams below, an appearance like a sea of blood.

> Letter from Berkshire County, Mass., Dec. 9, 1786, in *Connecticut Journal* (New London), Feb. 2, 1787, reprinted in *The Columbian Magazine* (Phila.), 1-7, March 1787.

The Jefferson Flood of March 1801

Twin disasters overwhelmed the "land of steady habits" in the middle and lower Connecticut River Valley in the March days of 1801. On the 4th, Thomas Jefferson was inaugurated as president of the United States,

223

thus ending the New England-backed rule of the Federalist Party in the nation's capital, and on the 18th and 19th of March a great rainstorm spread a record-making flood over the region. Partisan political editors lost no time connecting the dual calamities, and everafter the inundation was known locally as "the Jefferson Flood."

The *Vermont Journal* at Windsor carried a notice of flood damage in Connecticut, but made no mention of local losses. Two editors in Cheshire County across the Connecticut River in southwestern New Hampshire described the storm and flood on March 18 and 19, 1801:

> Much damage was done in this town and vicinity by the violent storm of Wednesday and Thursday last. The water rose higher than has been known for 20 years past, and we hear of a large number of mill-dams being carried away, and most of the bridges.
>
> *New Hampshire Sentinel* (Keene), March 21, 1801.

> Very copious and severe rains fell the last week in this town and vicinity. The damage done to public roads and private property cannot be accurately estimated at present, but is undoubtedly considerable. In Westmoreland [opposite Putney] the damage including that done to individuals, will amount to 3,000 dollars.
>
> *Farmer's Museum* (Walpole), March 24, 1801.

The Great Snowmelt Flood in Late April 1807

The weather turned warm after the prodigious snows in the first week of April 1807, though little rain fell. Soon the ice-locked rivers and snow-clad hills could not resist temperatures that ran as high as 78° at Montpelier. All became fluid and sought passage to the sea by the quickest route. The four or five feet of snow in open fields dissolved in a few days and a great flood was on.

On April 20 and 21, many bridges were taken out, and mills and buildings undermined and carried off. In the upper valley, bridges at Newbury across the Connecticut River and at Bath, New Hampshire, across the Ammonoosuc River were victims of the combined force of water and ice. The mills at Lebanon, New Hampshire, across from White River Junction were destroyed. At Newbury, a highwater mark was cut into a rock on Colonel Tenney's property, and this height was not attained again until the flood of 1876. Adino Brackett at Lancaster (N.H.) wrote in his diary: "Water has risen to an amazing height. I never perceived it so high before. The present height is accurately marked on two elms and a pine."

Some of the details of the destruction can be gleaned from contemporary newspaper reports:

Freshet. – The water in Connecticut River was never known to be higher than it now is. We had some considerable rain the week past, and being much snow on the mountains back, has raised the streams beyond the usual bounds at this season. Bridge damaged but still standing. Whirlpool caused by the damaged bridge undermined the toll house which dropped twelve feet, now held from going down the river by ropes.

Brattleboro Reporter, May 2, 1807.

Destruction by water. – By the sudden rise of the rivers on Monday, and Tuesday last, the following bridges, &c. have been destroyed – One bridge in Reading – two bridges over Quechee river in Woodstock – one bridge over White river in Royalton – one bridge over Onion river in Montpelier – one bridge over Connecticut river in Newbury, near Wells river – a saw mill on White river falls – and damage sustained by the canal at Water-Quechee Falls, $7,000.

Connecticut Courant (Hartford), May 6, 1807.

The Deluge of July 1811

"We have to announce the most extraordinary tidings of ruin and destruction, by the late heavy rains, from various quarters, which have ever occurred within the memory of the oldest inhabitants of the country," declared the editor of the Windsor *Vermont Journal*.

Within a few short hours on the morning of July 22, 1811, the countryside of Windsor and Rutland counties was turned into a sodden morass as a result of a deluge that released tons of water on every mile of hill and valley. Some of the physiography of the region was completely transformed by the torrents of water, the economic wealth of manufacturing and agricultural industries was greatly reduced, and death dealt to a number of individuals.

The quick-rising freshets on this July morning seem to have been equaled in magnitude in Vermont history only by the downpours on the afternoon of November 3, 1927. The latter was statewide, but the earlier storm seems to have concentrated its intensity over a smaller belt of the central part of the state, comprising Rutland and Windsor counties. Most likely a cold front, stalling over the area for a number of hours, triggered the repeated heavy showers.

The editor of the *Vermont Journal* at Windsor summarized the many reports of devastation that poured into his office in the week following the disastrous event:

> The late Freshet in some parts of the State, has been the most destructive ever known. A gentleman of respectability, just returned from the Westward has handed in the following hints.
>
> All the crops of Grain and Grass, on the low meadows of Otter Creek, which were never more promising, have been swept away or buried in ruins.
>
> At Sutherland's Falls, on Otter Creek, the Forges and Mills are all swept away.
>
> At Whitehall, all the buildings at the Falls—Saw Mills, Grist-Mills, Carding Machine, &c. &c. are gone.
>
> At Middletown, 11 buildings, including the Mills, &c. &c. have been carried away; and also,
>
> At Poultney, Maj. Topp's Woolen Factory, with all the stock, including 4,000 lbs of Wool. Two persons were drowned here.
>
> At Fair-Haven, were carried off all the Iron Works, erected by Col. Lyon, including Forges, Trip-hammers, slitting-mills, plating-mills, &c.
>
> In Clarendon, Capt. Parker's large two story House turned over, four houses were carried away and entirely destroyed, with various other Damage almost beyond calculation.
>
> On this side of the Mountain, it has also done much damage. Most of the bridges and mills on Quechee River, are swept away, and many on White River. The roads are cut up and destroyed in a most shocking manner. It is impossible, at present, for carriages to pass from this [Windsor] to Middlebury. Six miles of Judge Keyes's Turnpike from Stockbridge to Rutland, are said to be so totally destroyed, that it cannot be repaired, or rebuilt in the same place.

Some details of the storm's impact on Middletown, the Rutland County community that appeared to be the chief sufferer, were described by Reverend Lemuel Haynes in a communication to the *Vermont Gazette* at Bennington:

226

On Monday morning, the 22d last, and through the day, we were visited with uncommon heavy showers of rain from S.W. attended with thunder. According to different measurements of water caught in vessels standing distant from any buildings, the rain water fell from 12 to 15 inches perpendicular height. The effects were such as might be expected. Small rivulets, which had meandered through meadows, &c. and passed under fences without injuring them, now assumed aspects and powers of rivers, and swept all before them; changed their course, and cut channels like rivers, carrying away rocks, green trees, &c. High lands were surprisingly marred, and low lands deluged far beyond what the oldest men living had ever seen. Many acres of choice lands were ruined, part by being swept away to a great depth, and part by being overwhelmed to a considerable height with stones, gravel, trees, &c. Crops of all kinds, which in the morning have the most flattering prospects to the husbandman, before night were seen mingling in the common ruin. In low lands, the devastation was general. Although the rain began after the rising of the sun, yet it was so powerful that by nine or ten A.M. the streams began to be formidable. It was not long before bridges, mills, tan houses, yards, dwelling-houses, &c. went in rapid succession. A great part of the fields, meadows, &c. in town, had the fences swept away. . . .

On the whole, the thunder sounding, the rain falling, the waters roaring, banks caving in, houses undermining and sweeping away, men calling to each other for help, when little could be done, people fleeing their houses to save their lives, and consternation and dismay in all faces, rendered the scene truly awful.

The Torrent of 1830

The weather during the early summer of 1830 had been cold and wet, but the middle of July turned very hot and oppressive to introduce one of the warmest weeks ever known in the usually temperate hills and valleys of the Green Mountains. A thermometer at Middlesex in the central Winooski Valley rose to 90° and above on seven consecutive days. With humidities continuing high in the tropical air mass, residents suffered from the high temperature-humidity index and looked forward daily for relief in the form of showers. What came was more than they bargained for.

During the last week of July more than seven inches of rain dropped into Professor Zadock Thompson's rain gauge at Burlington. Over half this amount fell during a sixteen-hour period on the 26th, sending all the rivers and streams in western Vermont and northeastern New York into high flood stage.

Under the heading, "Awful Calamity," the *Middlebury American* gave details of the storm damage in Addison County:

227

After a week of extreme and oppressive heat, heavy rain storms commenced last Saturday evening, and continued almost incessantly, until yesterday noon. The streams just north and east of this place (in which directions the rain fell most rapidly), have been swollen to an unparalleled size. Most of the bridges have been swept away, and the crops are injured to a very great extent. Between this place and Vergennes, we learn that the tops of many fields of grain appeared just above the surface of the water.

Immense damage has been done to buildings, farms, crops &c. to the north and east of us—our account of which is necessarily imperfect. But all the desolation of other places is trifling compared with that which has swept through New Haven West Mills—the besom of destruction appears to have literally moved all but the naked rock where the cluster of houses stood, and left there the gloom of death.

The *Vermont Watchman* at Montpelier told of the flood damage in the vicinity:

In this village, two bridges on North Branch, the office of J.Y. Vail, Esq., a barner's shop and a wood shed destroyed. In the east part of Middlesex, a saw mill, an oil mill, grain mill, woolen factory, school house, barn, were swept away, and two dwelling-houses overturned. The bridge across the narrows, which stood 60 or 70 feet above low water mark, and the arch bridge below, are gone. Of the little village of Moretown, consisting of sixteen buildings, six only are left; all the mills except one, and most of the dams, together with a great number of dwelling-houses and barns on Mad River have been swept away. Not a bridge on Mad River is left standing. In Waterbury Village the water was 6 to 8 feet higher than ever known, and many of the houses were inundated and much property destroyed, no one anticipating danger.

The event was memorialized by one of its near-victims, Lemuel B. Eldridge, in: *The Torrent; or an account of a deluge occasioned by an unparalleled rise of the New-Haven River, in which nineteen persons were swept away, five of whom only escaped, July 26, 1830.* Eldridge described the destruction of property in Lincoln, Bristol, and New Haven where $31,000 worth of damage was done. At New Haven Hollow or West Mills, the flood waters swept the twelve members of the Stewart and Willson families, along with two neighbors, to eternity. Trapped on an island with the Stewarts and Willsons, Eldridge and his nineteen-year-old son sought safety on an improvised raft that was quickly swept downstream. Eldridge was soon dumped into the torrent but managed to cling to a bankside some three-quarters of a mile downstream, where he spent the rest of the night fighting for his life. His son, repeatedly struck by floating timbers, expired nearby.

Since a number of copies of this unique pamphlet survive, *The Torrent* apparently found a wide audience among those who enjoy reading about disasters and contemplating their meaning.

228

The High Flood in July 1850

"Heavy rain with a strong wind from the southeast raised the Connecticut River to a height not known at this season for forty years," reported the Windsor press in describing the effect of the storm of July 19, 1850. A tropical disturbance moving northeast caused high floods in the Washington, D.C., and Philadelphia areas. The path apparently led on a northeast track over central Pennsylvania and New York State, putting Vermont in a southeast flow of tropical air brought north from the Gulf Stream area of the North Atlantic. At Burlington the rainstorm was reported to have been the heaviest since the seven inches that fell in July 1830. A Montpelier measurement gave the amount in 1850 at five inches.

The *Vermont Watchman* at Montpelier appraised the flood situation throughout the state: "In the section of this country, it appears that North and East the flood was somewhat less than 1830. South and South-west it was much heavier, and this was therefore, taken altogether, the greatest storm and flood that has ever occurred here within the memory of man."

The recently constructed railroads of Vermont suffered their first of many encounters with mountain floods. The Vermont Central Railroad received damages estimated at $30,000. The Rutland Road had washouts between Ludlow and Bellows Falls. At Chelsea, nearly every road bridge for eight miles was carried away by high water. In the Newbury area, the waters covered all but two or three acres of the intervales in the ox-bow and was said to have been the highest since the "Great Freshet of October 1785."

The Great Snowmelt Flood in 1862

The memorable "Snow Crust of 1862" (*see* page 41), which helped to preserve the deep snow cover underneath through March and half of April, was a major contributor to the unique flood of that year. The winter snowfall had been very heavy in the upper Connecticut River Valley: Lunenburg measured a total of 147 inches, and Lancaster (N.H.) nearby, 143 inches, a large portion coming in late winter. In early March a heavy crust formed on the deep snow cover, strong enough to support a team of horses.

Temperature held the key to this vast locked-in hydrologic reservoir. On April 12, the first steady flow of warming winds from the south penetrated into the northern valley. The Strafford thermometer soared to afternoon readings of 49°, 53°, and 57°, and on the last day remained above freezing through the following night.

No significant rains fell at this time, but the springlike warmth began to transform the two to three feet of hard snow pack from solid to liquid form. The great snowmelt was on. From Canada to the Massachusetts border, the Connecticut River rose to heights never before experienced.

A letter to the *Country Gentleman* from J. W. Coburn of Springfield described the impact of the flood on an individual farm:

> But this hope was destined to be blasted, for on the 18th of April the Connecticut filled its banks and began rapidly to expand over the alluvions, and continued to swell out of its dimensions until midnight of the 19th, when it reached a maximum of 7 feet higher than known before within this century. So rapid was the rise of water at the farm of your humble servant, that sheep had to be thrown onto scaffolds by the hundred to escape drowning—300 were boated off to the hills on the 20th—nearly the same number barely escaped by driving off on the ridges that the water had left uncovered. In the night of the 19th, horses and cattle were taken out of stables after several feet of water had come in. But a few sheep were drowned; the most severe loss was in a large quantity of fleece wool, 600 bushels of oats, and corn in cribs. What was so recently considered a beautiful farm is a perfect wreck—fences all gone, some of the outbuildings undermined; sheep racks, lumber, rails, and all manner of rubbish, scattered in the greatest confusion over every part of the intervale; cellar and lowest rooms in rear part of house filled with water, mows of hay soaked 6 feet from ground.

Mary R. Cabot, the historian of Brattleboro, declared the flood height of 1862 equal to any previous high stage and described some of the local destruction in the *Annals of Brattleboro*:

> From records made in old diaries it is supposed that the island opposite this village was under water in 1770 and in 1785. Just how much damage was done at that time is not known though it could not have been very great, as the island was in a high state of cultivation when the waters of the Connecticut River submerged it in April, 1862, and reduced the area from twenty-two to eight acres.

The All-New England Rainstorm and Flood of 1869

The greatest total volume of rainfall ever to descend on the New England states occurred during the late morning and early afternoon of October 4, 1869. After twenty-four hours of moderate to heavy rains, deluges of over three inches were loosed on the central New England hills within three hours, about doubling the amounts that had already fallen. Only coastal New England did not share in the unprecedented downpours, and that region had an offshore hurricane to occupy its attention on this day.

230

In Vermont, precipitation for the storm period at nine rain-gauge stations averaged 4.35 inches. The greatest catches came at Castleton with 6.57 inches and Woodstock with 6.35 inches. In the north Burlington measured the least with 3.71 inches. The rainfall period extended from about sunrise of the 3rd to late afternoon of the 4th, although some light amounts were recorded as falling on the 5th.

What caused the unprecedented deluge? A weather map analysis is difficult since observers took only three observations a day and were not as alert to the significance of small windshifts as are modern observers. But sufficient barometric reports are available to trace the movement of a trough of low pressure, in which a cold front was imbedded, across New York State and New England during the daylight hours of the 4th. Tropical air had occupied East Coast areas for several days under the aegis of a southerly flow. Rain commenced in Virginia and Pennsylvania on the 2nd and gradually spread over most of New England by sunrise of the 3rd. Rainfall amounts were heavy to excessive on the night of the 2nd-3rd in Pennsylvania and during the day of the 3rd in Virginia. Wind shifts and barometric dips indicate a cold front was making slow progress eastward across the Middle Atlantic States. The front separating tropical air to the east and continental air to the west reached the Connecticut River Valley about noon on the 4th and was responsible for triggering the excessive showers during the morning and early afternoon.

But why the excessive amounts? As much as 12.35 inches were measured in northern Connecticut at an exposed ridge location. At the very time that the cold airstream was edging slowly eastward in its trough of low pressure across New England, a vigorous hurricane – destined to become famous as Saxby's Gale – was speeding northward in the coastal shipping lanes southeast of Long Island and south of Nantucket. The tropical intruder crossed Cape Cod waters in mid-afternoon on the 4th, smashed into eastern Maine several hours later, and caused unprecedented wind and tide damage in the Province of New Brunswick along the Bay of Fundy about midnight. Though weather maps are lacking, it is surmised that the circulation of this small but powerful hurricane brought northward the usual tropical air in its circulatory system. As a southeast current aloft, the hurricane airstream was forced to glide over the cold front and its attendant cold air mass at the surface over interior New England. Thus, the usual heavy showers attending a cold-front passage were reinforced by a vast supply of tropical moisture carried aloft from the far reaches of the subtropical Atlantic Ocean. This turned the showers into deluges.

Damage in the South at Rockingham

The damage done to the town was almost beyond compute. At Bartonsville the Williams river overflowed its banks and made a short cut across a bow, sweeping away the highway, portions of the railroad track, the depot and several dwellings. Bridges were lifted from their foundations and swept down the stream. The waters inundated the meadows and carried away a great many fields of corn. Saxton's river rose with a sudden fierceness that was terribly destructive to property. The village of Saxton's river suffered severely, the damage done to that place being estimated at $75,000. A wool-pulling establishment, belonging to L. C. Hubbard, was carried away, with all its contents. Mr. Hubbard's loss was estimated at about $15,000. Messers, J. A. Farnswerth, J. F. Alexander, and Benjamin Scofield lost heavily. A fine meadow farm a short distance below the village, belonging to a young man named Barber, was damaged to the amount of $5,000. The river left its channel and cut a new one the entire length of the meadow.

It was one of the most remarkable floods of the present century. A large number of lives were lost.

Abby Maria Hemenway, *The Vermont Historical Magazine*, 5, 498-99.

Damage in the North at St. Johnsbury

Flood of 1869. A storm of thirty hours' duration, October 2 and 3, proved the most disastrous ever known in the town. Passumpsic River at the Center Village was two and a half feet higher than any former record; streets and houses were flooded; railroad tracks were washed away; trains were stopped; for five days there was no mail from the south; on the fifth day Postmaster Fleetwood set out with a mail of 1600 letters which he proposed in some way to get delivered at White River Junction.

On Sleepers River the turbid waters were floating timbers, trees, logs, wagons, horse powers and endless other miscellany; soon the cry was heard: "the bridge is coming!" and like a duck on the water, came sailing down the lumber yard bridge; it broke thro the highway bridge, pitched over the dam, took out the next bridge with a crash and hurried on; for a time it was held at the foundry bridge till a broadside of the castings' shop struck it and all went down the stream. In the new brick engine house, the engine was submerged; east of this was the scale packing shop, originally the grist mill built by Joseph Fairbanks in 1815, and the only survivor of early times; under the force of the flood it tottered and fell with a terrific crash and was carried off. Startling events were following each other with fearful rapidity, while hundreds of men stood powerless to avert further calamity. The power of the waters was seen in the floating down stream of a 600 pound lot of iron gearing. The work of cleaning up next day was a sorry spectacle, tools and machinery full of mud which lay in places two feet deep on the floor. Loss was $50,000.

Edward T. Fairbanks, *The Town of St. Johnsbury, Vt*, 519.

THE VERMONT WEATHER BOOK

"The Great Freshet of March 1876"

The spring rise of the Connecticut River in 1876 over its northern reaches brought the highest flood waters ever experienced, according to the testimony of two local historians of the region.

> We now come to what was probably the greatest rise of water ever known in this section. The Connecticut River was so high that the whole valley looked like a great lake: Guildhall Village was nearly surrounded by water, while opposite (the Connecticut River) in Northumberland village (New Hampshire) the streets were covered in many places to a considerable depth, and very heavy damage was done to several of the streets: in some instances they were gulled out from 15 to 30 feet deep. Many thought that everything in the vicinity of the falls would surely be swept away, including the toll bridge, saw and grist mills, paper mill, straw shed, and several dwellings; but luckily, as soon as the water had passed the mark of its previous highest altitude, the great river seemed to feel satisfied, and slowly the waters receded until the danger was over. This was the great freshet of the spring of 1876.
>
> C. A. Bemis, *History of Guildhall, Vermont*, 113-14.

> In 1876, after this part of the Connecticut valley had been dry and people had begun planting, heavy rains and melting snow in the mountains, caused a flood in which the river rose about eight inches above the highest water-mark before reached. The water was six inches deep on the kitchen floor of the house in which Charles C. Scales now lives, on the Upper meadow.
>
> F. P. Wells, *History of Newbury, Vermont* (1902), 262.

The High Flood of April 1895

Two heavy rainstorms moved over the Green Mountains during the second week of April 1895, and their combined runoff from snowmelt and precipitation raised the rivers of Vermont to their highest levels since 1869. Both storm centers followed rather erratic paths in their approaches. The first crossed the state diagonally from southwest to northeast on April 9, and the second moved southeast from northern New York to Long Island Sound on the 14th where it stalled for a few hours. Both opened the Green Mountain region to a flow of maritime air from the Atlantic Ocean.

Only moderate rains fell on April 8-10 during the passage of the first storm, but they caused much snowmelt, saturated bare soil, and raised water levels in streams and ponds. The second storm brought much heavier precipitation. Northfield measured 1.90 inches within twenty-four hours on April 13-14, and the catch at Norwich came to 2.70 inches. North Conway, New Hampshire, at the foot of the Presidential Range in New Hampshire, reported 6.25 inches storm total, indicative of the heavier amounts falling on high elevations.

In assessing the comparative heights of floods on the Connecticut River, the *Bellow Falls Times* stated:

It is generally believed that the flood of 1862 was the greatest in history and the flood of 1895 ranked second. The printed accounts however, do not indicate that there was much difference between the flood of 1895 and 1869, at both times the tunnel was banked and the railroad tracks were covered with water.

Flood damage west of the mountains was also extensive according to an account in the *Rutland Herald* of April 15:

The floods in this section today have been the worst since 1869, and the damage reaches thousands of dollars.

The water began to recede this forenoon, and then marks of desolation appeared on every hand. Farms were flooded, bridges washed away, railroad tracks undermined, houses partially submerged and roads badly washed.

Central Vermont unable to run trains from Burlington since Sunday. Down sleeper waterbound at Proctor — passengers transferred to city in teams.

Track passable to Chester, but washout near Bellows Falls with 150 feet of track and a culvert entirely washed away.

All sleepers over Rutland division of Central Vermont cancelled.

Bennington and Rutland running between two long lakes, but not stopped.

Clarendon & Pittsfield suffered severely. Trains not running. Marble finishing shops closed. At Middlebury cellars flooded and roads washed out. Big log boom at New Haven broke and logs moving toward lake. Vergennes wharves flooded and some damage done. Mountain roads are badly washed and stages almost impossible to run.

The Great Flood of March 1913

One of the great rainstorms of the century crossed the Ohio Valley and northeastern states between March 24 and 27, 1913. The attendant flood took a high toll in property damage and loss of life, especially in Ohio.

The flood situation in the Upper Connecticut River Valley was described by W. W. Neifert, the official of the U.S. Weather Bureau at Hartford, Connecticut:

The mild rainy weather of March 20-22 reduced to water the small amount of snow and ice remaining in the woods and mountains of the extreme upper watershed. Under usual seasonal conditions this small quantity of snow and ice would not have resulted in great floods; but the fact that considerable frost yet remained in the ground and the thoroughly saturated condition of the soil, made conditions favorable for a rapid run-off. The small streams soon filled to overflowing, breaking up the ice and causing gorges. These conditions being augmented two days later by heavy rains, a sharp rise in the larger streams quickly followed. The ice gorges and overflowing streams caused a moderate amount of damage in the small streams at Montpelier, Barre, Lancaster, Littleton, Lyndonville, and St. Johnsbury, Vt. Moreover, by reason of the fact that the initial stages of the larger streams were relatively high, a large volume of water was soon sweeping down the upper Connecticut with irresistible force, inundating thousands of acres of land, and doing damage of such extent that it can not be accurately estimated in a monetary sense. The flood was due chiefly to heavy rain falling on thoroughly saturated soil, which was only partly free from frost, and consequently many of the washouts and landslides occurred on frost formations. The absence of the usual spring covering of snow with its considerable water content, surely minimized the losses which occurred. However, the flood over the upper valley was the greatest since 1869, while in the lower valley it brought the highest water since 1896 [1895?].

On the headwaters of the Connecticut, as at Wells River, Vt., the river was at a relatively high stage on the morning of the 25th. It rose 3 feet, viz, from 27 to 30 feet during that day, and this rise, in connection with the rains which had fallen over the watershed, was the first intimation of a flood throughout the course of the river in Massachusetts and Connecticut.

At White River Junction, 46 miles below Wells River, the Connecticut River at 8 a.m. March 25, stood at 14.6 feet, and at 4 p.m. of that date it had risen to 20.3 feet; it reached a maximum height of 30 feet at 9 p.m. of the 27th, thus overtopping all previous high records back to 1869. While no precise measurements of the stages of the river at White River Junction during the memorable floods of 1869 and 1862 are at hand, a reliable witness of both floods is authority for the statement that the 1913 flood exceeded both of the earlier floods.

The breaking of a big log boom at Sharon, Vt. (on White River), by which 2,500,000 to 3,000,000 feet of logs were set adrift, was the cause of the loss of the highway bridge at White River Junction, and the placing in jeopardy of the Boston & Maine Railway bridge across the White River.

Losses were estimated at $37,000 for the bridge at White River Junction and at $10,000 for the logs at Sharon. A man was drowned at Highgate Springs and a railroad employee lost his life when a train derailed into the river at East Putney.

The Vermont Flood of 1927

"The flood of November 3, 1927, was the greatest disaster in the history of our beautiful State," declared Governor John E. Weeks. These were words from the heart, for the hills and valleys of the land of his birth lay in a state of ruin, and Vermonters suffered a state of mental shock at the magnitude of their personal and property losses. Furthermore, Governor Weeks had just suffered a grievous personal bereavement when his close associate, Lieutenant Governor S. Hollister Jackson, met a tragic death when engulfed by a stream of water surging down an erstwhile roadway. His automobile was swept into a water hole where he drowned within sight of his home. At least eighty-three other Vermonters met a similar end that fateful afternoon and evening, and property damage ran into the multimillions.

The first drops of rain splattered down on southern Vermont about 9:00 P.M. on the evening of November 2 as a cold front—imbedded in a long north-south trough of low pressure—moved eastward across New York State. The rain spread north to the Canadian border by midnight but fell lightly, since little moisture was available in the air mass then present over Vermont. Northfield near the central part of the state recorded only one-third of an inch up to 4:00 A.M. on the morning of the 3rd, about the expected amount from the approach of a cold front at this season.

About daylight, however, a new element entered the situation when a moist airstream from the North Atlantic Ocean commenced to flow over the hills and valleys of New England. This airstream had been brought northward well offshore earlier in the week by a tropical storm whose progress to the northeast had been blocked by a large anticyclone over Newfoundland. Meantime, the low-pressure trough from the west developed a secondary center at its southern end along the Carolina coast on the night of the 2nd-3rd. This secondary center soon absorbed the remnant of the tropical storm and consolidated the latter's moist airstream into its developing cyclonic circulation, which directed the tropical air over New England as a massive southeasterly current with wind speeds reaching from 30 to 40 miles per hour.

When the air flow reached into Vermont, it was confronted by a formidable barrier: the highland massif of the range of the Berkshires, Taconics, and Green Mountains rising to about 4,000 feet in places, and the dense, cold airmass introduced into the Champlain and Hudson valleys by the cold front of the low-pressure trough. Its considerable depth was now super-imposed on the crest of the mountain barrier of western Vermont. In the clash of opposing airstreams of different thermal and moisture compositions, the lighter and warmer tropical air was forced to glide upward and over the denser and colder air in its path. The forced ascension resulted in rapid cooling, condensation of moisture into raindrops, and precipitation in a deluge of unprecedented proportions for Vermont.

The Vermont terrain of 1927 had been well prepared for the occurrence of an extreme hydrologic event. October rains throughout the state amounted to an average of about 150 percent of the normal. In northern sectors, especially, the departures from the October expectancies were excessive: Bloomfield measured 8.78 inches or 290 percent of normal; Chelsea 6.75 inches, or 248 percent; Northfield 5.54 inches, or 233 percent; and Rutland 7.37 inches, or 212 percent. The rainfall periods occurred on the 4th, 13th, and 19-20th, sufficiently spaced so that no flooding occurred, but the soil became so super-saturated that any more heavy precipitation would immediately run off into the streams and rivers.

During the late morning and afternoon hours of November 3, all Vermont rainfall records were broken, both for hourly intensity and for single-storm totals, as the moisture in the southeast airstream was wrung out by its forced rise over the twin barriers. It was estimated by a contemporary meteorologist that a cubic mile of solid water had been lifted from the surface of the Atlantic Ocean and deposited on the hills and valleys of the Green Mountain State. Brooks, streams, and rivers swelled, reached bankful quickly, then exceeded flood stage when the unprecedented inundations swept down the hill and mountain sides.

As the valley floors filled, venerable landmarks were submerged and disappeared. Bridges collapsed and were carried downstream, washed-out highways and railroads cut communications, and countless structures tottered on their undermined foundations or were swept along in a pile of debris. Many helpless victims went to a watery grave. In all, Vermont lost eighty-four residents during the tragic events of November 2-4, 1927, the greatest toll of a natural disaster in the state's history.

The intensity of the rainfall in the Green Mountains can be judged from the chart of the recording rain gauge at Northfield in a mountain valley south of Montpelier where a first-class Weather Bureau station was equipped with a tipping bucket gauge that registered each .01 inch.

After accumulating only 0.32 inch in six hours, heavy falls commenced at 4:00 A.M. on the 3rd, and in the next seven hours 1.65 inches fell at the rate of about 0.20 inch per hour. Soon after 11:00 A.M. the rate became excessive. In the nine hours from 11:00 A.M. to 8:00 P.M. a total of 4.24 inches fell at the rate of about 0.50 inch per hour. The Dog River through Northfield crested at 6:30 P.M. after seven hours of the intensive rate. Though substantial rains continued for another sixteen hours, nearly three inches more falling, the Dog River did not rise above the height achieved in the first rush of flood.

The largest rainfall amount for the storm, 9.65 inches, was reported at lofty Somerset (elevation 2,096 feet) near the border of Bennington and Windham counties. Rain gauges were not exposed at higher eleva-

237

tions, but catches must have exceeded 10 inches and may have reached as high as 15 inches on the highest mountain slopes.

The United States Geological Survey estimated that 457 square miles of Vermont received over nine inches of rainfall, 1,660 square miles over eight inches, 3,320 square miles over seven inches, and 5,530 square miles (or 58 percent of the state's total area), over six inches.

The greatest contribution to the flood height in the Connecticut River Valley came from the White River in central Vermont, which had a higher run-off per square mile in proportion to its drainage size than any other stream in New England. At its peak the White River was discharging at a rate of 140,000 second-feet of water on the morning of the 4th, while the Connecticut River to the north was carrying only about 8,000 second-feet. Fortunately, the crest of the upper Connecticut River was delayed and did not reach the confluence with the White River until the next day and was of only moderate proportions. The only noticeable effect upon the flood height in the lower Connecticut Valley was to prolong the falling stage (that is, the lowering of the river height).

At White River Junction, the Connecticut River rose five feet higher than the former record of March 1913, and at Bellows Falls, 6.6 feet higher.

Winooski River Valley

The flood proved most destructive in the Winooski River Valley from the Barre-Montpelier area to the falls at Winooski near Lake Champlain. One remarkable feature was the rapidity of the rises of the small streams and then the rivers during the late afternoon and evening of the 3rd. There was no time for warnings and preparation except in the lower Connecticut River Valley, and in many places no time to escape. Tragedy followed tragedy in such rapid succession that people were left stunned and helpless for a time, and the losses of life and property were staggering for such a comparatively small area. The known dead in Winooski Valley numbered forty-eight persons, and the total damage amounted to an estimated $12 million.

Montpelier reported a high-water mark of 16.5 feet as compared with a previous high of 13.5 feet. As the surface of the main business street lies only six or eight feet above the normal river level, the entire downtown area was submerged under eight to twelve feet of water. The scene was described by local resident R. F. Wells:

> The water began to flow into the streets of the capital city at about 4 o'clock Thursday afternoon, November 3, both the main stream of the Winooski River and its North Branch, which flows under State Street, having overflowed the banks. Automobiles were caught in the rising tide and had to be abandoned in the streets,

238

The great flood of 1927 was not the first or the last to visit the state though it was the focus of most flood stories. This photograph, taken from a glass plate negative, shows the intersection of State and Main Streets in Montpelier at the height of the flood of 1869. The printing office at the right of the picture still stands, but the larger brick building in the middle has been replaced by the entrance to East State Street.

while merchants and shoppers were obliged to seek refuge on the second floor of downtown buildings before 6 P.M., all chance of getting home having been cut off. The water rose three feet an hour at times, by actual measurement.

There were hundreds of people compelled to spend two nights in the upper floors of the business blocks in the center of the city, in some cases without food, but only one life was lost, Byron Nelson, a clerk in the Dwinell Hardware Company store, being swept out into the street when he went downstairs to close a door. The water stood twelve feet deep above the street level in Main and State Streets, flooding even the second floors of some of the lower buildings, reaching its greatest height about midnight on Thursday and falling only gradually during the day—Friday. Some prisoners of the flood were taken out in boats late Friday afternoon

239

The full force of raging Otter Creek in the flood of 1927 failed to dislodge the sturdy Proctor railroad station which had withstood earlier assaults including the flood of 1913.

The highway and trolley car bridges over Otter Creek in Center Rutland were subjected to tremendous pressure during the height of the 1927 flood. Earlier the Center Rutland railroad bridge over the falls had collapsed.

The 1927 flood caused such extensive road damage that travel in some places was officially restricted. This sign, posted in the middle of Montpelier's State Street says: "Notice *All cars going toward Middlesex must obtain pass. Motor Vehicles.*"

This is the kind of pass that was issued to travellers during the days immediately following the 1927 flood. J.A. Johnson had to have the approval of Charles T. Pierce, Commissioner of Motor Vehicles, before he could go from Middlesex Notch to Plainfield and "return to Montreal subject to road conditions."

and others not until Saturday morning, when the water was wholly out of the streets.

There was no electric light anywhere in the flooded area. Telephone and gas service was entirely cut off, also. With candles, kerosene lanterns and electric flashlights, the people watched the waters rise through Thursday night, not knowing when houses might be lifted from their foundations and washed away, as several of them were on Elm Street and its adjoining streets, while barns and garages and debris of all kinds went floating by in the raging current, bumping into stout structures with a crash that shook them. It was a night of terror.

Farther down the valley, the flood's impact at Waterbury was described:

The Winooski overflowed its banks during the afternoon of November 3. At five o'clock it was above the highest previous water mark within the memory of the oldest inhabitant. A little later the course of the river was diverted, due to the wreckage piled against the Duxbury bridge at the east end of the town, and the rushing torrent poured down the main street of the village. From seven until nine it rose at the rate of four feet an hour and from that time on, at about one foot an hour until it reached its highest point at 4 o'clock Friday morning, November 4. It is estimated to have been from 15 to 18 feet above any previous high-water mark on record.

Many abandoned their homes entirely. Twenty-three refugees spent the night in the tiny cabin of the Rev. Robert Devoy, high on the hillside overlooking the Winooski Valley. Scores of others were marooned in the upper stories of their homes and were rescued

241

by boats throughout the night and early morning. About 400 people were sheltered in the High School building which was on high ground with its heating plant unimpaired. Many were taken in at The Tavern and Waterbury Inn.

Stories and Pictures of the Vermont Flood, November 1927. Compiled by R.E. Atwood. Burlington, Vermont, 1927.

The Double Flood of March 1936

Vermont experienced a cold and snowy winter during 1935-36, and rough conditions continued through the end of February. But other years have seen heavier snow cover on the ground, greater accumulations of ice in the lakes and rivers, and deeper penetration of frost into the ground. Heavier rainstorms have occurred than those marking the pre-equinox days in 1936, and more sudden onset of spring conditions have come in March with higher temperatures and more rapid thaws. But the middle days of March 1936 brought all of these factors—heavy snow cover, ice in the rivers and lakes, and deep penetration of frost—into concerted action on the same dates, and their combined effect spelled disaster for the river valleys of New England. A slightly different timing or interplay of these features might have prevented or greatly mitigated the extreme hydrologic conditions that created New England's greatest flood experience.

Winter precipitation across northern New England in 1936 ran a little above normal, as did the season's supply of snow. On March 3 about a foot of new snow was added to the ground cover still remaining in its frozen state. Some mountain stations reported 40 inches depth on the ground, and in Vermont several valley stations measured more than 30 inches. The high water content of the snow pack held the key to future events; it contained the equivalent of 12 inches of liquid water in the White Mountains of New Hampshire, and up to 10 inches at Vermont locations. A great mass of frozen precipitation was stored on the hills and mountains awaiting spring conditions that would bring its release to start downhill on the long journey to the sea.

During the middle two weeks of March 1936 the northeastern states lay under the influence of a vast atmospheric trough of low pressure extending from the polar regions to the Gulf of Mexico. Four distinct storm systems formed within this zone and extended their influence across the mountains of northern New England. Two were of major and two of minor proportions. Each moved on a path that subjected the Green Mountains to rapidly rising temperatures attending warm rainfall.

The first major disturbance moved north from the South Atlantic coast on March 11-12 to central New York State and then translated its energy

242

to a secondary center near Long Island. Heavy rains, prolonged and intensified by the slow movement of the storm system, deluged all New England. Over six inches fell in the White Mountains of New Hampshire and over four inches in southern Vermont within a twenty-four-hour period. The thermometer at Burlington soared to 54° on the 13th.

A second major storm generated in the southern end of the vast atmospheric trough on the 16th. The center moved northeast to the Atlantic coast, then turned north on the 18th and crossed eastern Pennsylvania and New York State on the 19th. This proved an even greater rain producer for New England. Two-day totals in excess of 7.50 inches were reported in northern New Hampshire. Rain fell over northern New England on every day from the 16th to the 22nd.

Total rainfall amounts for the entire storm period at Vermont locations were: Newfane 9.05 inches, Somerset 8.87, Wilmington 8.16, Searsburg Station 8.14, Brattleboro 8.02, and Cavendish 7.89.

The influx of warm air attending the first storm caused a typical spring ice flood with the thick covering on lakes and streams breaking up and flowing downstream. The heavy rains on March 11-12, augmented by the snowmelt runoff, caused a tremendous volume of water to pour into ice-clogged channels. Notable ice jams backed up the water in many places and raised the level of local flooding substantially.

The runoff from the still-melting snow cover and the excessive rainfall of the second storm reached river systems that were already burdened with the high waters of the first flood. A comparison of the discharges of the two floods in cubic feet per second from the Connecticut River illustrates the magnitude of the second storm:

	13th	*19th*
South Newbury	31,400	77,800
White River Junction	46,500	120,000
Vernon	109,000	176,000

Record high-water marks were reached at most locations in the upper Connecticut River Valley. Beginning in the vicinity of Fifteen Mile Falls above Barnet and extending south to the Massachusetts border, all previous known flood discharges were exceeded (except in the part of the river near White River Junction where the peak was less than in November 1927). The relatively low contribution of the White River at the time the upper Connecticut River peak arrived at White River Junction was the primary reason for not reaching a record stage there.

At the narrows near Bellows Falls, a crisis arose. With both roller gates on the dam raised as high as possible and flash boards off at the peak of the flood, about 29 feet of water passed over the top of the dam and 24 feet over the sills at the flashboard openings. Blocking the railroad tunnel alongside the dam and power plant saved many buildings and streets in downtown Bellows Falls from serious damage. The Boston and Maine railroad bridge immediately below the dam at Bellows Falls, although loaded with gondola cars as a precaution, was severely battered by ice, trees, houses, and barns that came driving down the river.

At Vernon Dam close to the Massachusetts line, the important electric power complex underwent a siege of peril when the spillwater below the dam rose about 36 feet above normal and submerged the top of the dam by nearly 11 feet. Instead of a normal fall of 35 feet over the dam, the drop was only eight feet. The power plant was completely flooded and put out of operation.

The Hurricane Flood of September 1938

The devastation caused by the passage of the Hurricane of September 1938 through New England cannot be fully appreciated without a consideration of the weather situation during the preceding week of heavy rains.

On September 12 and 13, a southerly current of warm moist air of tropical origin commenced to flow northward over the Ohio Valley, New York State, and New England as the result of the influence of a low-pressure system moving eastward over the Great Lakes. Moderate-to-heavy rains fell across Vermont, especially on the 13th when Rutland received showers totaling 2.37 inches and Woodstock, 1.13 inches.

Again on the 16th, another disturbance from the Midwest traveled east-northeast well to the north of the Green Mountain region, a route that reinforced the southerly flow of moist air and caused light to moderate rain: Somerset 0.31 inch and Bloomfield 0.33 inch.

The next storm period, covering five days from Saturday afternoon to Wednesday morning, September 17 to 21, had almost continual overcast and rain. During the twenty-four hours ending at 6:00 P.M. on the 19th, heavy showers measuring up to two inches fell over the south while less than one inch covered the extreme north. Intermittent showers continued for another twenty-four hours into the 20th, at rates exceeding those of the previous day, with individual station amounts ranging up to three inches.

The first precipitation associated with the approaching hurricane circulation from the south reached southern Vermont soon after noon on the 21st and spread rapidly north. Amounts attributable to the hurricane averaged from three to five inches. In the north, Burlington had 2.85 inches and Northfield 2.95 inches on the 21st alone. Stations in the Connecticut River Valley in central and southern portions of the state had more: Woodstock 4.45 inches and Brattleboro 5.50 inches. The five-day totals amounted to large figures, such as Brattleboro's catch of 8.22

inches and Somerset's 8.20 inches. Newport in the extreme north on the Canadian border reported the least, 1.87 inches. Typical storm totals were 4.10 inches at Burlington and 4.59 inches at Northfield.

The greatest damage from the hurricane deluges occurred at middle elevations along the smaller tributaries of the larger rivers draining the area. Windham County communities in the southeast, where the rainfall was greatest, lost roads and bridges and were out of communication with their neighbors for from three to five days. The *Brattleboro Daily Reformer* put together a pictorial booklet soon after the storm in which the onslaught of rain and flood were vividly described:

Flood Damage 1938

> *Brattleboro:* Whetstone brook, a raging turbulent stream, went over its 1936 level and inundated Flat, Elm, Frost, and Williams streets and sections of West Brattleboro, forcing several families to flee from their homes. Some were carried to safety by rescue squads of police and volunteers.
>
> And, keeping pace with their tributaries, Connecticut and West rivers rose steadily and dangerously, flooding all low areas within their reach. The crest of the flood on these streams was reached on Thursday morning [Sept. 22] after a great amount of damage had been done along the West River and Vernon roads.
>
> The surging Whetstone carried out a long iron bridge on Williams street and a smaller span leading into the Holton & Martin Lumber Co. yards. It left its true course and covered fields and lawns, spilling into cellars and stores in the inundated areas.
>
> *Wilmington:* Wilmington, that is the center of the town, presented a sorry sight on the morning of Sept. 22, when residents viewed the handiwork of the Deerfield river on its wild rampage the night before.
>
> Selectmen estimated that damage to town and state-aid roads alone would exceed $225,000, exclusive of route 9. Personal loss to property owners and residents of the river bank section was not estimated.
>
> The river, always a threat in flood times, surpassed all previous performances as the many small tributaries, reaching unprecedented heights, swelled the larger stream to overflowing almost without warning.
>
> Proud citizens found that the fine cement bridge in the center of town, built only four years ago, had dropped down on the west side, smashing the water main and cutting off the town's water supply. Bridges in all directions were also badly damaged or washed away and gaping holes in roads everywhere testified to the fierceness of the flooded streams and torrents of rain.
>
> *West Dover:* West Dover reported flood losses including among other bridges its covered bridge in the village, as well as a lot of

the road between it and Wilmington. Bridges that weren't destroyed were isolated by crumbling approaches. Water was said to be twice as high as in 1927.

Numerous homes were weakened by the rush of waters beneath them, and damage was more severe to barns and garages. William Harris' garage was swept downstream.

Londonderry: Londonderry families were evacuated from all but six homes, many rescued in canoes in their night clothing. In the north village half of the Williams brothers' mill was destroyed, and 75,000 feet of lumber carried off by the errant waters. Some of the logs were believed responsible for the destruction of the J. B. Johnson house in South Londonderry. The family escaped before their home fell into the river.

The water ran deep in River street, pouring in and out of first story windows of some houses. One end of the cement bridge in the center of the village was left out of reach of the highway. Few houses remained without some damage done to them.

Jamaica: Jamaica's damage exceeded that of the 1927 flood. All but two bridges on the road from Pike's Falls to Jamaica village were swept away, besides the steel bridge on Depot street, the road in that section, and Leon Cheney's mill. The river ran from three to five feet deep in the "back" street.

The river appropriated highways, lumber yards, and fields over which to run forcefully against everything in its path. Small dams were carried out. When Cheney's mill struck the cement bridge, it collapsed like an eggshell. Farther downstream it took with it a barn holding 25 to 30 tons of hay.

Flash Flood at Rutland in June 1947

Continued heavy rains over the Rutland area on June 2-3, 1947, caused the collapse of a power dam on East Creek in East Pittsford, some six miles north of Rutland, and the consequent flash flood down the valley did considerable damage. The release of the impounded waters flooded the lower portion of the city to an estimated depth of 10 to 15 feet, causing the evacuation of 500 families and the suspension of power plant output.

The mainline trestle bridge of the Rutland Railroad over East Creek collapsed, and several old wooden bridges on highways were destroyed. A large amount of meadow and cultivated ground along East Creek was buried under a thick layer of silt and gravel.

Damage in the Rutland area was estimated at $5 million. Other areas throughout the state suffered flooded highways and delayed traffic.

Rain fell heavily in the mountains east of Rutland. Rochester reported 3.76 inches on the 3rd, in addition to 0.34 inch on the 2d. Rutland's catch

246

The flood of 1947 did serious damage in the Rutland area. The remains of what was once a cement bridge south of West Street can be seen at the left of the picture. Photo courtesy University of Vermont Special Collections.

totaled 3.08 inches. Over the state the heaviest amounts covered the northeast where Barnet received 4.12 inches in twenty-four hours and a storm total of 4.53 inches.

The Damaging Flood of June 1952

A series of heavy showers in the last ten days of May 1952 culminated in very heavy downpours on May 31 and June 1. Since the soil was well saturated, the streams in the northern part of the state rose to the highest levels since the great flood of March 1936.

The climatologist for New England, Mark T. Nesmith, surveyed the situation in Vermont in his monthly report:

> On the 1st, torrential showers produced flash floods, mainly in New Hampshire and Vermont. Streams rose rapidly during the afternoon and night. Highways were blocked by water or by land-slides; communications were disrupted; power and light services

247

suspended in many communities. The district most seriously affected was north-central Vermont, where two to four inches of rain fell within a few hours on already saturated valley soils and high stream levels. Many families were removed from flood-threatened dwellings in Orange, Caledonia, Washington, and Orleans counties. A score of towns were partially evacuated; ten major highways closed; fifteen bridges washed out. At Ludlow in Windsor Co., the American Red Cross reported sixty homes damaged by flood waters of the Black River. In Montpelier, a gate on the Winooski River was dynamited to prevent a major inundation of the business area.

Rainfall over Vermont during a twenty-one hour period from May 31 and June 1 averaged close to 3.50 inches. The greatest amount was reported at Plymouth in the central part of the state where 6.01 inches fell, but no other station approached this amount, even at high elevations. Cavendish reported 4.23 inches, Reading Hill 4.14 inches, Dorset 4.03 inches, and South Londonderry four inches. In the upper Winooski watershed, amounts ranged over three inches: Montpelier Airport 3.68 inches, Northfield 3.48 inches, and Barre 3.16 inches.

The state's capital experienced the highest waters since the great March Flood of 1936. In the northeast, water ran four feet deep in the streets of Orleans and St. Johnsbury, and all electricity was knocked out in North Danville. Damage throughout the state was estimated at $500,000, with the bulk reported from the Barre-Montpelier area. A landslide occurred at Lincoln. Four persons lost their lives during the flood. Three drowned when a car went into the high waters near Cavendish.

Flash Floods in June 1960

Heavy rains of a local cloudburst type hit a triangular section of south-central Vermont during the night of June 3-4, 1960. The towns of Ludlow, Ascutney, and Chester lay in the center of the excessive precipitation zone. The Black and Williams rivers rose quickly after the sudden downpours, reaching heights not witnessed since the Flood of 1927, and soon overflowed their banks. A house near the center of Ludlow washed away when flood waters poured down the main street. Thirty families were evacuated. A dam and other river installations received severe damage.

The greatest amount of rainfall at regular measuring stations, 3.77 inches, occurred at Cavendish on the Black River above Ludlow. No other station reported half that amount, the highest being 1.56 inches at Peru and 1.30 inches at Somerset, both high elevations.

High Water in June 1973

Heavy rain began on Thursday night, June 28, 1973, and continued into Saturday morning June 30, before tapering off to scattered, but still locally heavy showers. In some places the worst flooding since 1927 resulted, causing Vermont to be declared a disaster area by President Nixon. Total damage reached an estimated $64 million.

The cause of the heavy rains was a stationary front extending through New York State, which forced a strong southeast flow of moist oceanic air over the Green Mountains. Several stations reported over 6 inches in forty-eight hours, with maximums of 7.50 inches at Jamaica and 7.19 inches at South Londonderry, both in the south-central section.

Damage to state highways amounted to an estimated $10 million. Considerable crop losses resulted. The rampaging Black River, while invading the heart of downtown Ludlow, wrecked the General Electric Company plant.

A six-year-old boy drowned when he fell into the overflowing Branch River at Bennington, and a seventeen year old boy drowned in the Black River near Cavendish.

Hurricane Belle's Flooding in August 1976

The first hurricane since 1960 to threaten the New England coastal area lost its full strength in the cool waters south of Long Island before making a landfall in the early morning hours of August 10, 1976. Belle continued northward as a vigorous extra-tropical storm carrying a huge load of tropical moisture. It moved on a north-northeast track the length of New Hampshire about parallel to the Connecticut River.

Vermont, in the western sector of the storm, received heavy deluges. As much as four inches fell over the Green Mountains on the 10th: Readsboro 4.11, Mt. Mansfield 4, Ludlow 3.97, and Whitingham 3.60 inches.

Winds of gale force accompanying the storm downed power lines and trees. Many roads and bridges washed out in flash floods that also destroyed crops and deposited gravel and silt over fields. Several mobile homes were swept into streams and ruined. Two deaths and one serious injury occurred.

HIGHLIGHTS OF THE WEATHER, 1986–JANUARY 1994

F or the eight years covered in this summary, 1986 to 1993, Vermont's usually variable weather can be described only as variable. Monthly average temperatures for the state ranged from 69.8°F in the hot July of 1988 down to 5°F in the all-time coldest December of 1989–the coldest month of any name since the coordinated record-keeping program for the state began in 1895. The monthly temperature variations within each year canceled one another out, however, and the annual average differed only 2.9°–from a low of 41.2°F in 1992 to a high of 44.1°F in 1991. The absolute extreme temperature readings for the period ranged from 99°F in August 1988 down to −43°F in February 1993 (with an unofficial press report of −46° at Craftsbury). Neither of these readings broke records: the all-time low for Vermont remains −50°F at Bloomfield on December 30, 1933, and the high remains 104°F at Woodstock on July 21, 1916, and at Cornwall on August 21, 1916.

Precipitation totals during 1986–1993 were above normal in four years and below normal in four. Monthly precipitation amounts ranged from only 0.52 inch in an unusually dry February 1987 to 7.05 inches in August 1990. The latter came at the end of a wet summer that had higher than usual amounts of rain from May through August.

Monthly precipitation amounts also showed trends. In 1990 only two months had a deficiency, resulting in an annual total of 52.83 inches (129 percent of normal). At the other end of the rain gauge, 1988 was consistently dry, ending with a total of only 35.81 inches (88 percent of normal).

Unlike the snowy winters of the 1960s and 1970s, the last five years of the 1980s were not memorable for deep snows. Readings taken at Northfield, a typical valley station in the north-central part of the state, reflected a seasonal snowfall average of 80.0 inches for the period 1986 to 1993. The annual totals ranged from 51.9 inches in 1988–89 to 98.3 inches in 1986–87. January 1987 was the snowiest month throughout the state, with several valley stations accumulating over 50 inches during the month. The maximum snow depth on the ground, excluding mountain top stations, was 58 inches, measured at West Danville in the northeast in March 1993. The winter of 1992–93 was notably snowy, the entire state averaging more than 30 inches in both February and March 1993.

The heaviest snowstorm of this period dropped more than 30 inches on November 12–14, 1990, in Orleans County. The most severe winter storm occurred on March 13–14, 1993, during the Blizzard of '93, which piled up 31 inches at Waitsfield and 29 inches at Brookfield.

Threat of Hurricane Gloria in September 1985

For the first time since 1938, the Green Mountain State was threatened with the overhead passage of an intense storm from the tropics when Hurricane Gloria crossed central Long Island, hit the Connecticut shore with the low barometric pressure of 28.50 inches (965 millibars), and headed north up the Connecticut River Valley. High pressure over Quebec caused the center of the disturbance to take a path to the northeast rather than to the northwest as did the New England Hurricane of 1938. Vermont therefore received very heavy rain but not the high winds associated with hurricanes. Chittenden measured 3.63 inches of rain, Rutland 3.60 inches, Cavendish 3.44 inches, and Woodstock 3.32 inches. Widespread stream flooding resulted in Bennington, Rutland, and Windham counties.

Early Heavy Snowstorm in October 1987

The principal weather event of 1987 was the record early snowstorm on October 3–4. Totals up to 20 inches were reported in the central and southern parts of the state. This exceeded the former record for early snowfall in October 1836. Amounts on valley floors and along the Connecticut River were less than in the high terrain of southwestern Vermont. Because trees were still in leaf, they suffered severe damage.

251

Downed utility poles and tangled wires caused widespread outages in more than 30,000 homes.

Hot Summer of 1988

The warmest July since 1959 and the warmest August since 1984 combined to give Vermont a very hot summer in 1988. These months were 2.0°F and 2.1°F above normal. The hot, humid conditions started in late June and continued through mid-August. Maximum readings in July soared to 99°F at Bellows Falls and Vernon in the southeast, and to 98°F at St. Johnsbury in the northeast. Vernon had fourteen days with 90°F or above in July and fifteen such days in August. The extreme was 97°F at Burlington on the 3rd. Dryness in May and June were corrected by above normal precipitation: statewide averages were 4.11 inches in July and 5.59 inches in August.

Coldest December 1989

The outstanding event of recent years came in December 1989 with the coldest month of any name in the Vermont records since 1895. The statewide average of 5.0°F was 16.7°F below normal. This was 4.4°F lower than the previous coldest December in 1917 and 0.8°F lower than the previous coldest month, February 1934. On twenty-eight of thirty-one days, a reading of zero or below was registered somewhere in the state. The remainder of the winter averaged above normal for January and February.

July 1990 Flood in Northern Vermont

On July 5 heavy rain from thunderstorms produced rapid and intense flooding across northern Vermont. Rainfall amounts from 3 to 6 inches over 2 to 3 hours were common. Most small streams in Caledonia, Lamoille, Washington, and Franklin counties overflowed; many roads and a few bridges were washed out.

Main stem river flooding was confined in the Lamoille Valley between Stowe and Moscow. In the headwaters of the Little River, a campground was devastated by several feet of water.

At North Montpelier high water caused a portion of a dam's embankment to fail. As a result, huge sections of Factory Street were washed out and the hydroelectric plant was shut down.

Flash Floods in August 1990

Sharon, Windsor County

Three to five inches of rain fell in the area beginning August 5. On the 7th a beaver pond is suspected of having overflowed. The spilling caused a washout and the wreck of a Central Vermont Railroad freight train.

Worcester, Washington County

The North Branch of the Winooski overflowed onto Route 12 during the afternoon of August 7, 1990, forcing the road to be closed for a short time. Minister Brook also caused some localized flooding during the same period.

Stormy November 1990

High winds on the 3rd brought down trees, power lines, and a few utility poles and blew off roofs. High winds were blamed for starting a brush fire when downed electrical wires touched off some leaves. Several homes and a car were hit by falling trees, motorcycles were overturned, and dumpsters were blown about parking lots.

A storm system started as rain on November 12 and changed over to snow by the 14th. Accumulations were reported in excess of 30 inches in Orleans County. The heavy, wet snow took down power lines, causing outages that lasted up to ninety-six hours in some parts of the state. The combination of cold and power loss forced some homeowners to drain their water pipes to prevent freezing.

Winds gusted up to 51 mi/h at Montpelier. Higher gusts were reported in other parts of the state but could not be confirmed by official measurements. Four duck hunters had to be rescued from Lake Champlain when their 14-foot boat proved no match for 16-foot waves. They managed to make Sunset Island, where they were rescued by a Coast Guard helicopter.

Stormy March 1992

Snow and freezing rain fell across parts of northern Vermont on March 1, 1992, ahead of a warm front moving over northern New York State. Freezing rain that led to black ice was followed by 2 to 5 inches of snow, creating traffic hazards. Waitsfield received the most snow, with 5 inches on the ground by dark.

253

Freezing rain ahead of a warm front in the Mid-Atlantic region overspread northern Vermont on March 6. Back roads became dangerously icy.

Ice Jam Floods of March 1992

A strong disturbance crossed from southeast New York through central New England on March 11, dumping 0.5 to 1.5 inches of rain across Vermont. The rain, combining with runoff from snowmelt, caused ice in the rivers and streams to start moving. Ice jams resulted in extensive flooding of lowlands. In Montpelier alone most of the downtown area was inundated with 4 to 6 feet of water when ice jammed along the Winooski River under the Bailey Avenue bridge; seventy-five homes and 118 businesses sustained water damage. The power of the jam also knocked out an iron railroad bridge in the center of the city, twisting the steel rails and splintering the wood ties.

On March 27 a low-pressure system moved across northern New York while a second center formed along the New Jersey coast and moved to the waters off Cape Cod by the morning of the 28th. Between 0.5 and 2 inches of rain fell across Vermont. The runoff, combined with more snowmelt, caused ice in the streams and rivers to move and led to another ice jam flood. Portions of Route 110 in Orange County went under 3 to 4 feet of water, the Waits River flooded Route 25 in Bradford, the White River inundated East Barre and Tunbridge, the Browns River cover Route 128 between Westford and Essex Center. The Winooski flooded at several places in Chittenden County near Richmond. Route 100 in Londonderry was closed for an interval.

"Crown of Winter Storm" on March 28–29, 1992

In New England a late-season storm, usually toward the end of March, that raises the snow cover to the greatest depth of the season is called a "crown of winter storm." Late on March 28, 1992, a deep-pressure system moved into the Gulf of Maine, dumping 6 to 12 inches of new snow across north-central and northeastern Vermont and 4 to 6 inches across the remainder of the state. The totals included 11 inches at Island Pond and East Haven, 10 inches at Newport and Canaan.

Severe Thunderstorms in September 1992

Severe thunderstorms moved across much of Vermont ahead of a strong cold front in the afternoon and evening of September 10. Hardest hit was the Albany area, where a 40-by-24-foot garage was lifted off its foundation and thrown 60 feet, a barn had a 40-foot section of its roof torn off, and a pickup truck was damaged. In parts of Franklin County, thunderstorm winds downed trees and power lines and took the roof off a mobile home in Highgate, pushing the home off its foundation. In Franklin eight cows were electrocuted when lightning struck the tree under which they were standing.

Cold February 1993

For the New England region, February 1993 was the third coldest on record, exceeded only by February 1923 and February 1934. In Vermont the average temperature of 10.4°F, or 8.4°F below normal, was the sixth coldest February, surpassed by February 1907, 1914, 1923, 1934, and 1979. Newport, on the Canadian border, had no reading above freezing during the entire month of February 1993. The minimum figures throughout the state came on February 7th, with the following official readings: Island Pond −43°F; Enosburg Falls −38°F; Marshfield −35°F; West Danville −35°F. The press carried unofficial reports of −46°F at Craftsbury, −44°F at Morgan, and −43°F at Canaan.

Snowfall that month was heavy, averaging 33.8 inches throughout the state. Amounts were greatest on mountain tops. Mt. Mansfield counted 64.6 inches total snowfall, which measured 69 inches on the ground by the 25th. Jay Peak had a depth of 66 inches on the 24th. At lower elevations, West Danville reported 44 inches on the ground on the 24th after receiving 42 inches of new snow during the month.

Wintry March 1993

The weather of March 1993 proved exceptional in all New England and was the second month in a row that was below normal in temperature and above normal in snowfall. The Northeast averaged 4.1°F colder than usual and western Vermont a raw 4.5°F colder. The greatest station departure from normal was 4.7°F at Rutland. The lowest minimum reached −25°F at Island Pond on March 19.

255

Snowfall in all sections averaged slightly over 30 inches, well above normal. The greatest snowfall total was 59 inches at Jay Peak, where 80 inches lay on the ground on the 16th. Mt. Mansfield counted 93 inches on the ground on the 21st, the greatest in Vermont for many years. At lower elevations, South Lincoln at 2,020 feet above sea level reported 51.4 inches new snowfall and 38 inches on the ground on the 15th. West Danville at an elevation of 1,575 feet measured 58 inches depth on the 15th. Vermont shared with most of the East in the Blizzard of '93 on March 12–13, when West Burke reported 31 inches of new snow and Brookfield 29 inches. Island Pond's 21 inches of fresh snow raised the total on the ground to 52 inches at the end of the big storm. The severe conditions temporarily shut down the Grand Isle ferry.

Cold January 1994

January 1994, with an average of 5.9°F, went into the Vermont record books, tying for the coldest January since coordinated records began in the state in 1895. It also tied for the third coldest month of any name, exceeded only by December 1989 (5.0°F) and February 1934 (5.8°F) in the ninety-nine-year record.

The absolute minimum of −38°F in January 1994 was registered at six locations: Canaan, Essex County; Enosburg Falls, Franklin County; Mt. Mansfield, Lamoille County; Huntington, Chittenden County; Searsburg, Windham County; and Union Village, Orange County.

Following the passage of a low-pressure system with its attendant cold front down the St. Lawrence Valley in the early morning of December 26, 1993, the way was open for the almost continual free flow of frigid arctic air from the northern Hudson Bay region into northern New England. The temperature at Burlington International Airport dropped to −14°F the morning of December 27 to begin the intense cold period that witnessed only one overnight thaw until near the end of January. Burlington experienced twenty-one days with a minimum temperature of zero or below during January, including a nine-day stretch of mornings of zero or below from January 15 to 23. On only two days did the thermometer get above freezing in January until the 29th, when it zoomed to 46°F after being at −29°F only thirty-six hours before. The lowest reading during the month came with −29°F on January 27, 1994, one degree shy of the record all-time minimum for the Burlington station. Montpelier began the morning of the 27th at −30°F and saw 50°F by the afternoon of the 28th.

January 1994 was also the third snowiest January on record at Burlington. At 38.6 inches, this was one storm shy of the existing January record of 42.4 inches set in 1978 and below the 41.3 inches in January 1966. (The all-time snowiest month at Burlington was 56.7 inches in the cold December of 1970).

WEATHER
WATCHING
IN VERMONT

WEATHER OBSERVERS

Everybody must have talked about the weather conditions experienced during the first decades of settlement in Vermont, but no one was able to make a precise statement about its extremes without the availability of scientific measuring instruments. Thermometers were a rare item in colonial America since they had to be procured in Great Britain, shipped across an often stormy ocean, and transported inland by wagon over rough and rocky roads. When the *Connecticut Courant* ran a listing of January temperature readings at Hartford during the Hard Winter of 1780, the editor felt compelled to append a note explaining what a thermometer was, how it functioned, and what its figures meant. If one can judge from the temperature readings noted in the newspapers of the day and in personal diaries, about a score of thermometers existed in the colonies at the outbreak of the Revolution.

No references to local thermometer readings appeared in Vermont newspapers until 1790. On a brief visit during the summer of 1786, James Winthrop, the librarian of Harvard College, brought a thermometer and a barometer to Lake Champlain in order to make a survey of its elevation, but the instruments were not employed in meteorological work.

Several localities were favored by having among their early settlers an individual who took a special interest in scientific matters and started a local record of the weather. These were men working on their own initiative without institutional or government support.

At Rutland

The Republic of Vermont was fortunate to have among its early settlers one of the leading academic scientists in North America. Reverend Samuel Williams (b. Waltham, Mass., April 24, 1743 O.S.; d. Burlington, Vt., January 2, 1817) served as Hollis Professor of Mathematics and Natural History at Harvard College from 1780 to 1788. He was the third incumbent of this post, having succeeded his former instructor, the distinguished John Winthrop VI. From the latter Williams probably acquired his interest in meteorological studies since Winthrop maintained a valuable series of daily observations at Cambridge from 1742 until his death in 1779. Williams continued the meteorological program at Harvard and widened its scope by exchanging data with the Meteorological Society of the Palatinate at Mannheim, Germany.

Early in 1788, Williams ran into financial and personal embarrassments. He was reputed to have led a gay social life, being addicted to wearing bright-colored jackets in contrast to the drab black in style among most of his colleagues. His wife was fond of feminine finery. Even worse in the eyes of some, both were said to engage in drinking expensive wines.

Williams apparently was hard put to support this style of life. It was soon discovered that he had misappropriated 286 pounds from Harvard's Hopkins Fund and also falsified a signature on a note. Further irregularities came to light in his carrying out of the duties of executor of the estate of a fellow clergyman. Williams was indicted by the Massachusetts General Court, and this led to the appointment of a faculty committee to inquire into reports of his questionable conduct that might "reflect dishonor upon the University."

Appraised by a friend of the adverse sentiment of the committee, Williams submitted a hurried letter of resignation, which was accepted. With haste he departed the Cambridge scene and the jurisdiction of the Commonwealth of Massachusetts. Harvard's loss was Vermont's gain since Williams chose to move his residence from the United States of America to the independent Republic of Vermont where he would be free from prosecution. At this time the Green Mountains served as a frontier refuge for many souls who found its free air preferable to the confining atmosphere of the older settlements.

Williams settled in the village of Rutland in the summer of 1788, and for the next thirty years pursued a useful life as preacher, editor, publisher, historian, and college lecturer, as well as serving as a keen scientific observer of the developing region. As to the character of the man and the esteem in which he was held, one of his fellow citizens of Rutland published a brief sketch in 1797:

> Of Samuel Williams, L.L.D. Member of the Meteorological Society, in Germany, of the Philosophical Society in Philadelphia, and

262

Rev. Samuel Williams, clergyman, scientist, author, and editor, lived from 1799 until his death in 1817 in Rutland where, among many other undertakings, he founded The Rutland Herald. *He hastily left Harvard College in 1788, where for eight years he had been Hollis Professor of Mathematics and Natural History, when it was discovered he had misappropriated Harvard funds. The independent country of Vermont, obviously beyond the jurisdiction of Massachusetts courts, proved hospitable, and Williams continued making his daily meteorological observations. In 1794 he wrote and published* The Natural and Civil History of Vermont *containing "the best-rounded analysis of weather and climate of any American locality."*

of the Academy of Arts and Sciences in Massachusetts, it may with propriety be said, that "he is the most enlightened man in the State in every branch of Philosophy and Polite Learning; and it is doing him no more than justice to say, there are very few in the United States possessed of greater abilities, or more extensive information: added to which, he is a most excellent orator, and always speaks in a manner best adapted to the understanding and capacity of those whom he addresses. In the year 1794 the Doctor wrote and published the *Natural History of Vermont*, executed much to his honour, and the great satisfaction of all Naturalists. In politeness, ease, and elegance of manners, Doctor Williams is not inferior to the most polished European Gentleman.

Williams's observing equipment included a thermometer, barometer, and rain gauge. He does not appear to have possessed a registering (i.e., maximum/minimum) thermometer, but employed his sunrise and 3:00 P.M. observations as the maximum and minimum readings. He also noted the direction of the wind along with cloud and sky conditions. Of particular interest to him was the temperature of wells at different depths, a subject then drawing the attention of his colleagues throughout the country.

In 1794, Williams published his major work, *The Natural and Civil History of Vermont*, in two volumes containing 416 pages. It was the most ambitious study of a special region of the United States yet published at that time and contained the best-rounded analysis of the weather and climate of any American locality. One chapter is entitled: "Climate—An account of the temperature, winds, rain, snow, and weather. The change of climate which has attended the cultivation of the country."

Williams employed an analysis of his meteorological observations at Rutland for 1789, 1790, and 1791 to establish the climatic extremes for that part of the Green Mountain region in order to make comparisons with observations made in other parts of America and Europe. He attempted to account for the change of climate that had allegedly taken place since the first settlement of the country by Europeans, attributing the change to the cutting down of trees and the cultivation of open fields where once dense forests stood. Along with the climate section of Thomas Jefferson's *Notes on the State of Virginia (1781)*, Williams's contribution stands in the forefront of early American regional studies.

At Bennington

A thermometer was located at Bennington at least as early as 1790 since local temperature readings appeared in the *Vermont Gazette* that year. The instrument may have belonged to a member of the Philosophical Society of Bennington whose meetings were mentioned in the press during the decade of the 1780s. Just to the south over the Massachusetts border in Williamstown, meteorological instruments were in use at Williams College as early as 1796, though no continuous record was preserved until 1811.

At Windsor

A series of records including temperature, wind conditions, and rainfall were kept at Windsor at least as early as 1801 since the results appeared from time to time in the local *Vermont Journal*. Attributed to B. Fowler, they were republished in 1806 and 1807 in the *Boston Medical and Agricultural Reporter*, whose editor made an early attempt to collect weather reports from various New England localities so that regional comparisons could be made.

At Burlington

While serving as the first president and one-man faculty of the University of Vermont in the opening years of the nineteenth century, Reverend Daniel Sanders maintained a set of weather records at Burlington for several years. He had been schooled at Harvard, Class of 1788, and most likely studied science under Professor Samuel Williams. Only monthly means of temperature for the years 1803 to 1808 have survived, thanks to their appearance in Zadock Thompson's *History of Vermont*, published in 1842. The detailed original records have not been located.

At Middlebury

A resident of Middlebury possessed a thermometer as early as 1806 when a local reference to a high reading during a notoriously hot summer appeared in the local newspaper. Regular observations at Middlebury College become available in 1811 when monthly summaries appeared in the local *Literary and Philosophical Repository*. This was edited by Professor Frederick Hall, a science enthusiast, who collected parallel observations from Yale College in Connecticut, Williams College in Massachusetts, and Bowdoin College in Maine. Their publication in the magazine through 1816 constituted a second attempt to make a collective of New England observations. No original manuscript records have been found in the Middlebury College archives or in the library of the Sheldon Museum.

265

At Norwich

A valuable meteorological record was provided by Captain Alden Partridge, U.S. Army, when he spent a year visiting his old home at Norwich across the river from Hanover, New Hampshire. His observations cover the months from July 1811 to April 1812, spanning the "Hard Winter of 1812," reputedly the severest since the "Hard Winter of 1780." His thermometer at Norwich dropped to −31° during February 1812 and did not rise above 35° in that cold month. The data were published in the *New American Medical Register*, which served as the leading exchange for scientific information of the day. Records made by Partridge at West Point, New York, before and after the Vermont interlude were also included in the published article. Captain Partridge later returned to Vermont to found and head the Military Institute at Norwich, but no record of further weather observations have been found.

At Hanover, New Hampshire

Weather observations made on the campus of Dartmouth College at Hanover provide a valuable indicator of general conditions prevailing over nearby parts of Vermont. Thermometer reports for as early as 1810 appeared in the *Dartmouth Gazette* when a reading of −17° was mentioned on the famous Cold Friday of that year. No continuous records appear to have been maintained until the 1820s when Professor Ebenezer Adams took up the task. His manuscripts have been preserved in the Dartmouth Archives.

A series of regular meteorological observations began in 1834 and have been continued to the present with very few breaks, making it one of the most venerable and valuable of New England records. Professors Ira and Albert Young were associated with the observational program during the 1850s and supplied valuable historical material for the region in their contributions to the *Dartmouth Gazette* and the *American Journal of Science*. Some Hanover summaries appeared from time to time in the *Vermont Chronicle*, published in Windsor as the organ of the Congregational Church of Vermont.

At Newfane

Martin Field engaged in weather observing as a scientific hobby at Fayetteville (now Newfane) for many years. From November 1826 through April 1833, his analysis of a series of observations and summaries were published in the *American Journal of Science*, the prestigious publication of Professor Benjamin Silliman of Yale College.

Field was a graduate of Williams College, Class of 1798, and received an honorary degree from Dartmouth College in 1805. After studying law at Chester, he settled at Fayetteville in 1800 and entered upon a fruitful career. "He was eminently successful in his profession, and for nearly thirty years enjoyed a large and lucrative practice, which he was compelled to abandon by reason of his excessive deafness. On relinquishing his practice, he commenced the study of Geology and Minerology, and by great perseverance and industry collected what, at that time, was regarded as the rarest and most extensive cabinet of minerals in the State," in the words of Abby Maria Hemenway.

Field served for ten years as state's attorney for Windham County and on several occasions represented his town in the General Assembly and at constitutional conventions. In 1819 he was elected major general of the first division of the Vermont militia.

Field's record consisted of three temperature observations per day, and the maximum and minimum. He also noted the weather character of the day, as well as the prevailing wind direction. He had a rain gauge and noted the depth of each snowfall. A column was reserved for the number of days with lightning and thunder and another for aurora borealis occurrences.

At Huntington

Weather observing in Vermont during the middle years of the nineteenth century was greatly enriched by the unique contributions of James Johns, who was the son of the first settler of Huntington. From this vantage point high on the slopes of Camel's Hump, Johns began a "Journal of the Weather" in 1830; all but five years of the span from 1830 to 1873 have been preserved in the archives of the Vermont Historical Society. Apparently, Johns never owned a thermometer or other instrument, though he often copied data from other localities in his daily journal.

Johns is well known in bibliopolic circles as the "Vermont Pen Printer," for his creation of the "Huntington Gazette," which he composed and printed with pen and ink, much in the manner of a medieval scribe. He also showed native originality in producing stories and poetry with a local slant. Some of his poems dealt with the more unfavorable aspects of Green Mountain weather, such as backward springs, cold summers, and long winters, whose occurrence he facetiously attributed to evil spirits.

267

At Newbury

A worthwhile series of non-instrumental observations began in 1823 on the Coos intervale at Newbury in the upper Connecticut River Valley. David Johnson, a son of one of the earliest settlers of the region, kept his eye on the sky while tending a general store and farming the rich lands along the river. His property today occupies a scenic site on Route 5, just north of the village of Newbury. Johnson usually made three observations a day of wind and sky conditions and summarized the number of days each month that different conditions prevailed. He secured a thermometer in 1835 and started making snowfall measurements in 1852. Johnson's observations covered a span of forty years, at least to 1862.

Along with many regular reports of the weather-observing network of the local academies throughout New York State, Johnson's record after 1852 was published in the *Annual Reports of the New York Board of Regents* as an example of conditions in northern New England. Stations with records of varying length for eastern New York, such as those at Plattsburgh, Salem, Cambridge, Troy, and Albany in the path of Vermont's prevailing westerly airstream, make New York State data helpful in studying the course of Green Mountain weather events in the pre-Civil War period. Johnson also kept a diary of phenological matters including the budding and flowering of trees, the arrival and departure of birds, along with observations of the aurora and comets.

At Randolph

William Nutting maintained a weather diary at Randolph from 1828 to 1849 and again in 1863. It is assumed that the manuscript books for the long gap from 1849 to 1863 have been lost. The record is descriptive, except for thermometer readings from 1828 to 1831 and again from 1844 to 1849. Nutting was especially interested in snowfall and usually made note of the daily amounts and totaled these into monthly and annual amounts. His snowfall data fill a void in Vermont snowfall records that is especially useful during the years of the big snows during the 1840s.

After graduation from Dartmouth College in 1807, Nutting became principal of the newly established Orange County Grammar School. While continuing to teach, he studied law in the office of Judge Dudley Chase, the leading jurist of the area. At this time Nutting was offered the post of chairman of the department of mathematics and natural philosophy at the University of Vermont. Though fond of teaching, he declined in favor of pursuing the practice of law among his neighbors.

Nutting served as justice of the local court for twenty-three years and as town clerk for nineteen years. He also continued his interest in the local school system as secretary and treasurer of its board of trustees for many years.

Though bedridden toward the end, Nutting continued to record his weather observations, usually two entries per day in his own hand, until his last birthday about four weeks before he died. On November 26, 1863, his daughter wrote in his weather diary: "Father died at 12½ o'clock M, aged 84 years 27 days it being the day of national thanksgiving."

At Burlington

"Of all the persons who have written concerning Vermont, probably no one has equalled Zadock Thompson in time and labor expended and in useful publications produced," wrote historian Walter H. Crockett in 1921.

Throughout his life, Thompson eked out a bare living from his positions as writer, educator, clergyman, and state official. He was always

Despite ill health which hindered many of his projects, Zadock Thompson wrote a number of books including his Natural History of Vermont, *first published in 1853, and still considered a valuable reference. Despite his industry, he and his family were frequently poverty stricken.*

Zadock Thompson was a scientist as well as a geographer, historian, writer, and teacher. For many years he kept careful meteorological records at his Burlington home. His wife assisted him, as clearly shown in this chart for January, 1856. Note the "Remark" for the 19th of the month, "Mr. Thompson died 5 p.m."

the teacher, whether from the pulpit, the lecture platform, the schoolroom desk, or the printed page. He devoted his scholarly talents to investigating and analyzing the various aspects of natural history, concentrating in the fields of geology, astronomy, and meteorology.

As a student, Thompson worked his way through the University of Vermont by producing a Vermont almanac known as "Thompson's Almanac," after its editor, publisher, and salesman. Within a year after graduation in 1823, he produced a comprehensive *Gazetteer of Vermont*, and followed this with a *History of Vermont from Its Early Settlement to the Close of the Year 1832*. After a decade of further research, the original work was expanded into the *Civil, Natural and Statistical History of Vermont*, which contained 656 closely printed, double column, octavo pages. The section of this work entitled "Climate and Meteorology" includes discussions of local temperature, winds, rain, snow, seasons, smoky atmosphere, dark days, Indian summer, meteors, earthquakes, aurora borealis, magnetic variation, remarkable seasons, and a comparative view of climate. Summary tables of Thompson's own observations at Burlington were included. These covered temperature, barometric pressure, and precipitation, along with many comments of current weather events. His original manuscript books found their way to the Smithsonian collection of early American weather records at Washington where they may be consulted in the National Archives.

After her husband's death, Mrs. Phoebe Thompson carried on his observing program for the remainder of 1856. No Burlington reports have been preserved at the National Archives for 1857, but on January 1, 1858, Professor M. K. Petty began another series for the Smithsonian Institution, which continued through 1864.

At Lunenburg

Hiram Cutting spent most of his life in remote Essex County. He was one of those people of native talent who was able to overcome the intellectual isolation of a small community and achieve prominence in his fields of science, medicine, and education. When almost fifty years of age, he was awarded an M.D. from Dartmouth College for his medical researches, and later received an honorary A.M. and Ph.D. from Norwich Institute. Cutting is said to have belonged to no less than eighty-nine scientific, literary, and medical societies, either as an honorary or paying member.

When a youthful school principal of sixteen years of age, Cutting undertook a weather-observing program, and with the occasional assistance of family and friends he maintained the record for Lunenburg with only small breaks until his death in 1892. He forwarded his monthly reports

271

Vermont's scientists have not always been products of academic institutions. Hiram A. Cutting of Lunenburg, Vermont, had little formal training but his studies in the fields of medicine and the natural sciences brought him honorary master's and doctor's degrees. His "Meteorological Tables and Climatology of Vermont, with Map Showing the Rainfall; also, Suggestions and Directions about Foretelling Storms" was published in 1877 while he was serving as state geologist.

to the Smithsonian Institution, the U. S. Army Signal Service, and finally to the New England Meteorological Society, covering forty-four years in all.

While serving as state geologist in 1877, Cutting published a summary of his first thirty years of weather watching in a pamphlet entitled *Meteorological Tables and Climatology of Vermont, with Map Showing the Rainfall; also, Suggestions and Directions about Foretelling Storms.* He was also author of a "Natural History of Essex County," which was published in Hemenway's *Vermont Historical Magazine.*

The Essex County records were continued after 1892 by other observation stations, first at Stratford, New Hampshire, and later at Bloomfield, a short distance upriver on the Vermont side. Thus, the Northeast Kingdom possesses a continuous sampling of its stimulating weather types for a period of more than 120 years. With the closing of the Bloomfield station in 1968, the observer at West Burke in nearby Caledonia County continued the regional reports.

Wilson Alwyn Bentley

"Jericho, Vermont, has one industry, if such it may be called, that gives it a unique place and that has carried the name of the town all over the world." With justifiable pride, the co-editors of *The History of Jericho, Vermont* described their debt to Wilson Bentley, "The Snowflake Man" of Jericho, who became known to thousands of readers of popular technical and photographic magazines during the opening years of the twentieth century. Yet several decades passed before the academic profession in the United States and throughout the world came to appreciate the significance of Bentley's investigations of the physical nature of precipitation that he carried on at his farm in a remote location in northern Vermont where he was born and died. Today his pioneer researches into the mysteries of raindrop and snowflake formation have accorded him the title of "America's First Cloud Physicist."

When still a youth, Bentley took his first photomicrograph of a snowflake during a storm on January 15, 1885. For the next thirteen years,

"Snowflake" Bentley of Jericho stands by his special camera used for photographing snowflakes and frost traces. Wilson A. Bentley was interested in the many manifestations of the weather, but his special concern was the snowflake. His studies and pictures made clear that no two snowflakes are alike.

while working quietly by himself, he took about 400 photomicrographs of ice crystals. Through the interest of Professor George Perkins of the University of Vermont, the first article describing and illustrating his work appeared in 1898 in Appleton's *Popular Scientific Monthly*. During the seven years from 1898 to 1904, Bentley made 344 measurements of the size of raindrops by a method of his own invention, in addition to carrying on his work with snowflakes. In 1904, the results of his work were published in the *Monthly Weather Review* of the United States Weather Bureau. The article was described as "an incredible paper which on the basis of ingenuity and number of new ideas is perhaps unmatched in the world's scientific writings on raindrops and raindrop phenomena" by his recent biographer, Professor Duncan Blanchard of the State University of New York at Albany.*

In 1924, the first research grant ever awarded by the American Meteorological Society was given to Bentley for "40 years of extremely patient work." Bentley was never much concerned with money or its acquisition. In 1926, this was made clear with his statement: "From the practical standpoint I suppose I would be considered a failure. It has cost me $15,999 in time and materials to do the work and I have received less than $4,000 for it." The meager income came mostly from payments for articles and from the sale of slides of his photomicrographs, mainly to schools and museums.

Though recognition of his genius came late in life, Bentley's name is now known throughout the scientific world as a result of the publication of *Snow Crystals* in 1931 by the McGraw-Hill Book Company of New York City. Nearly 2,500 photographs of the some 4,500 that Bentley collected were reproduced. A new paperback edition was issued in 1962, so great was the demand not only in scientific circles, but by all interested in the delicate patterns of nature's beautiful art.

The Smithsonian Institution Period

The middle years of the nineteenth century witnessed a successful attempt by a federal institution, other than the military, to organize a national weather reporting system. Professor James P. Espy, nationally known as the "Storm King" because he lectured frequently about storms and was known for his vitriolic attacks on his critics, was appointed in 1842 to a clerkship in the Navy Department. This was arranged by his congressional supporters so he could collect and analyze weather reports and develop his theories about storm behavior.

In 1843, Espy listed three Vermont correspondents who forwarded

*Weatherwise.

monthly summaries of their daily weather observations: A. Putnam of Grafton, John Hunt of Bennington, and J. C. Bryant of Cambridge. Contributions also came from the Connecticut River Valley communities of Keene and Charlestown in New Hampshire. In the following year a report from a Middlebury observer expanded the Vermont data in Espy's growing files.

The proposed national weather observing system did not materialize at this time, but the experience gained was put to good use in 1848 when the newly established Smithsonian Institution, under the leadership of Professor Joseph Henry, sought to create a weather reporting system as one of its initial activities. During the first year of operation, reports were received from Brattleboro, Burlington, and Montpelier. The Capital City correspondent was Daniel Thompson, the Vermont novelist, whose *History of Montpelier* (1860) contained a wealth of interesting local weather data and lore.

The Smithsonian weather reporting program was explained to Vermonters by F. Holbrook of Brattleboro in one of his regular columns in *The Cultivator*, a weekly farm magazine published at Albany, New York, and widely read by progressive farmers of the time. The observers were supplied with a thermometer and rain gauge and requested to make three temperature readings a day—7:00 A.M., 2:00 P.M., and 9:00 P.M. At these hours the sky condition and direction and estimated speed of the wind were also noted. Daily rainfall totals were recorded. A column for comments often included interesting details about storms and other weather phenomena, and some observers enriched their reports with details of phenological activity. Entries were made on a prepared form and forwarded to Washington by franked mail at the conclusion of each month.

The Smithsonian network expanded in the early 1850s, continued to function on a reduced scale during the difficult years of the Civil War, and carried on its program through the end of 1873. At least forty-five sets of Vermont records of varying quality and length were forwarded to Washington, providing good documentation of many local weather events and climate trends for twenty-five years.

Some records were of a fragmentary nature, extending only a few months or having missing data, but several dedicated observers maintained valuable records over a considerable period of years. Among the most useful were those gathered at:

Brandon, 1852-69	Middlebury, 1851-52; 1865-70
Burlington, 1849-64; 1871-73*	Norwich, 1856-59; 1871-73*
Castleton, 1852-55; 1869-73*	Randolph, 1865-73
Charlotte, 1868-73*	St. Johnsbury, 1853-55; 1857-59
Craftsbury, 1854-73*	Woodstock, 1868-73*
Lunenburg, 1859-73*	

*Record was continued under the direction of the U. S. Army Signal Service after 1873.

United States Army Signal Service: 1871-1892

The need for a nationwide storm-warning service was brought to public attention by a series of marine disasters on the Great Lakes and along the Atlantic coast during the autumn and early winter of 1869. After much bureaucratic wrangling in Washington over what agency would be in control, Congress passed a joint resolution on February 4, 1870, calling for the establishment of a weather reporting and storm-warning service; and five days later President U. S. Grant assigned the task to the U. S. Army Signal Service since it was the most experienced governmental organization in the field of communications.

With two organizations engaged in meteorological work, the Smithsonian Institution agreed to transfer the supervision and support of its climatological observers to the Signal Service at the end of 1873. Interested mainly in storm-warning activities, the Signal Service neglected the volunteer observers across the nation, and many soon ceased to forward their monthly reports to Washington. A handful of Vermont observers remained faithful to their daily weather observing, so some information about local weather events exists during what has been called the "dark age" of American climatology. During this period, valuable records continued at Burlington, Charlotte, Lunenburg, Newport, Strafford, and Woodstock.

The shipping on Lake Champlain was considered of sufficient importance to establish a first-class weather observing and forecast relay station at Burlington in 1871. A national telegraphic circuit carried the Burlington report to Washington three times daily where it was entered on the national weather map. Forecasts for the Lake Champlain area were distributed to navigation interests. The station and its services fell victim to one of the periodic economy drives that sweep Washington from time to time and was closed in 1883.

New England Meteorological Society: 1884-1896

When petitioned by Congress to serve the agricultural interests of the country in better fashion, the Signal Service sought to revive the climatological network by encouraging the faithful of the Smithsonian era to continue their observations and by enlisting new contributors. A plan was formulated to encourage individual states to establish their own weather services, following the initiative of Iowa in 1878. Eight other states complied by 1881 in the plan that envisaged at least one observer in each county of every state.

None of the New England states established its own service at this time. Rather, a group of individuals connected with educational institutions or government agencies in the Boston-Providence area organized the New England Meteorological Society in June 1884, and *Bulletin No. 1*, containing a summary of the observations, was issued. Individual observers throughout New England were invited to participate and send their observations to the editor of the *Bulletin*. Four Vermont observers were included in the first report:

Burlington:	W. B. Gates	Newport:	Rev. E. P. Wild
Charlotte:	Miss M. E. Wing	Pittsford:	L. Woods

By January 1886, the number of Vermont observers increased to thirteen, sufficient to supply representative averages for the state. Temperature and precipitation data have been compiled for each year since 1886 to establish the basis for statewide normals.

United States Weather Bureau

The quasi-military aspects of the federal weather service drew frequent public criticism during the 1880s, and several scandals involving the handling of public funds increased the demand for the creation of a civilian agency to carry on the scientific work of meteorology. Accordingly, a Weather Bureau under the aegis of the United States Department of Agriculture was created in 1891. All meteorological work of the military was transferred to the new agency in March 1892.

The Weather Bureau took over the supervision of the collection and publication of the climatological data in cooperation with the New England Meteorological Society. After several years, the latter was dissolved and the New England network became a section of the nationwide system of the federal government. Volume 1, Number 1 of *Climate and Crops, New England Section* in February 1896 included the following Vermont and Connecticut River Valley observers:

Brattleboro: W. H. Childs	*Jacksonville:* Miss M. French
Burlington: W. B. Gates	*Northfield:* U.S. Weather Bureau
Chelsea: George Dewey	*St. Johnsbury:* Fairbanks Museum
Cornwall: C. H. Lane	*Strafford:* H. F. J. Schribner
Enosburg Falls: J. H. Mears	*Wells River:* R. E. Pember
Hartland: E. A. English	*Woodstock:* H. F. Dunham
Irasburg: O. W. Locke	

Known as cooperative observers, the volunteers are equipped with an instrument shelter, a set of self-registering maximum-minimum thermometers, and a rain gauge. They maintain daily records of temperature, precipitation, prevailing wind direction, sky condition, and weather of the day, as well as noting unusual natural phenomena. The system has continued with minor changes to the present, though the supervising agency became the Department of Commerce in 1940. The former Weather Bureau is now known as the National Weather Service, an agency of the National Oceanic and Atmospheric Administration.

Vermont has had two first-order weather stations carrying on both observing and forecasting functions. Burlington was one of the original Signal Service stations, being established on May 27, 1871, in the United States Customs House and continuing near the same location until June 15, 1883. In October of that year, a cooperative observer, W. B. Gates, began a series of observations on a more limited scale and continued through March 1906, providing a useful record of more than twenty-two years. In 1906 a new first-class station was established by the Weather Bureau in a building at 601 Main Street, which it purchased for the purpose. Observations were carried on at this location near the University of Vermont campus until June 1943 when all operations were moved to the Municipal Airport in South Burlington, expanding the limited operations started there in 1939. The station now operates from the north wing of the Administration Building of the Airport.

A second first-order station was opened in 1887 by the Signal Service on the grounds of the Norwich Military Institute, then located at Northfield. It occupied quarters in the main building with instruments exposed on the roof. The station continued in operation until 1943 when it was closed because the location did not directly serve wartime aviation interests. A cooperative station has been maintained at Northfield almost continuously since to preserve the continuity of this valuable record for north-central Vermont.

At the present time, weather observations are taken at a number of airports throughout the state and transmitted on a national teletype circuit. Hourly observations are made at Burlington, Montpelier, Rutland, and St. Johnsbury, and either three-hourly or six-hourly at Newport and Wilmington. Similar stations near the border of Vermont supply airport stations at Lebanon and Keene in New Hampshire, Glens Falls and Plattsburgh in New York, and Sherbrooke in Quebec.

As of May 1996 there were forty-six cooperative stations.

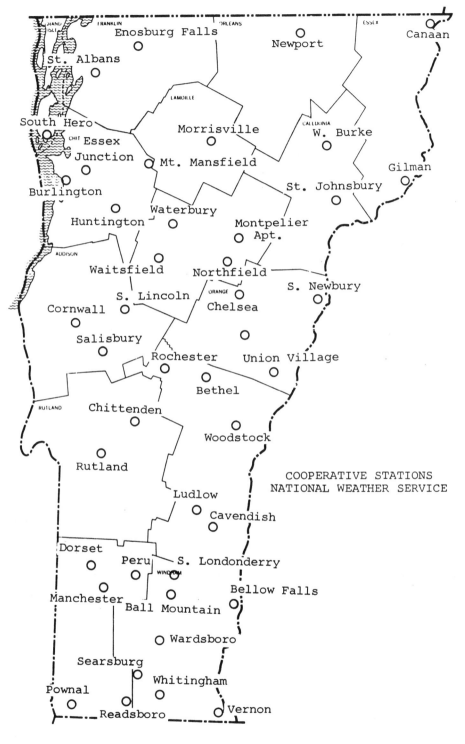

Enosburg Falls ○

Newport ○

Canaan ○

St. Albans ○

Morrisville ○

W. Burke ○

South Hero ○

Essex Junction ○

Mt. Mansfield ○

St. Johnsbury ○

Gilman ○

Burlington ○

Waterbury ○

Huntington ○

Montpelier Apt. ○

Waitsfield ○

Northfield ○

S. Newbury ○

S. Lincoln ○

Chelsea ○

Cornwall ○

Salisbury ○

Rochester ○

Union Village ○

Bethel ○

Chittenden ○

Woodstock ○

Rutland ○

COOPERATIVE STATIONS
NATIONAL WEATHER SERVICE

Ludlow ○

Cavendish ○

Dorset ○

Peru ○

S. Londonderry

Bellow Falls ○

Manchester ○

Ball Mountain ○

Wardsboro ○

Searsburg ○

Whitingham ○

Pownal ○

Readsboro ○

Vernon ○

279

WEATHER OBSERVERS

Burlington	Chittenden	National Weather Service
Canaan	Essex	Frederick W. Cowan
Cavendish	Windsor	Central Vt. Public Service
Chelsea	Orange	Gordon T. Heath
Chittenden	Rutland	Edmund H. Davenport
Cornwall	Addison	Malcolm B. Harding III
Danville	Caledonia	Otis F. Brickett
Dorset	Bennington	Charles B. Gilbert
Enosburg Falls	Franklin	O'Shea Publishing Co.
Essex Junction	Chittenden	William P. Hall
Gilman	Essex	Georgia-Pacific Corp.
Huntington Center	Chittenden	Mrs. Maude G. Shattuck
Ludlow	Windsor	Mrs. Laurel Tucker
Manchester	Bennington	Austin Fox
Montpelier	Washington	FAA - Knapp Airport
Morrisville	Lamoille	Inactive since 9/30/81
Mount Mansfield	Lamoille	Mount Mansfield TV Corp.
Newport	Orleans	Mrs. Pearl T. Drew
North Danville	Caledonia	Inactive since 12/17/79
Northfield	Washington	William E. Osgood
Peru	Bennington	Charles Black
Pownal	Bennington	Mrs. Ruth Falkner
Readsboro	Bennington	New England Power Co.
Rochester	Windsor	Rodney L. Johnson
Rutland	Rutland	Mrs. George Kirk
St. Albans	Franklin	Radio St. Albans Inc.
St. Johnsbury	Caledonia	Fairbanks Museum
Salisbury	Addison	Salisbury Fish Hatchery
Searsburg Station	Windham	New England Power Co.
South Hero	Grand Isle	Ray W. Allen
South Lincoln	Addison	Capt. Harry F. Wiseman
South Londonderry	Windham	Inactive since 5/1/84
South Newbury	Orange	Mrs. Isabel Whitney
Union Village Dam	Orange	U.S. Corps of Engineers
Vernon	Windham	New England Power Co.
Waitsfield	Washington	Harvey Horner
Wardsboro	Windham	Dr. Courtney C. Bishop
Waterbury	Washington	David R. Mac Lean
West Burke	Caledonia	Gilman W. Ford
Whitingham	Windham	New England Power Co.
Woodstock	Windsor	William R. Gould Sr.

TEMPERATURE
AND PRECIPITATION
RECORDS

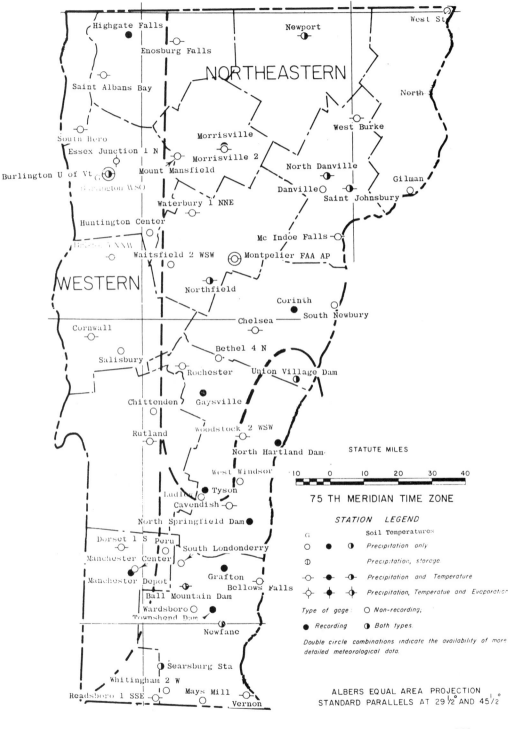

STATUTE MILES

-10   0   10   20   30   40

75 TH MERIDIAN TIME ZONE

STATION LEGEND

G   Soil Temperatures
○   ●   ◑   Precipitation only
◍   Precipitation, storage.
-○- -●- -◑- Precipitation and Temperature
-○̊- -●̊- -◑̊- Precipitation, Temperatue and Evaporation

Type of gage: ○ *Non-recording;*
● *Recording* ◑ *Both types.*

Double circle combinations indicate the availability of more detailed meteorological data.

ALBERS EQUAL AREA PROJECTION
STANDARD PARALLELS AT 29½° AND 45½°

TEMPERATURE AND PRECIPITATION RECORDS

	JAN.	FEB.	MAR.	APR.	MAY	JUNE	JULY
1895	18.4 (2.92)	16.3 (2.09)	26.4 (2.35)	41.3 (2.95)	57.5 (2.81)	66.9 (3.14)	66.7 (2.87)
1896	15.2 (1.35)	17.4 (4.69)	22.9 (8.12)	43.5 (1.20)	57.9 (1.71)	61.9 (2.41)	69.1 (3.85)
1897	19.5 (4.08)	21.6 (2.50)	30.2 (4.50)	46.4 (3.75)	59.0 (5.89)	59.9 (7.66)	70.9 (9.26)
1898	15.8 (6.07)	22.6 (4.53)	34.8 (1.39)	40.4 (3.30)	53.7 (4.16)	64.0 (4.10)	69.2 (2.36)
1899	16.5 (2.69)	16.9 (2.90)	24.9 (7.11)	41.9 (1.67)	54.4 (1.69)	65.3 (3.33)	67.7 (4.81)
1900	18.3 (4.48)	18.1 (5.80)	20.9 (5.11)	41.7 (1.36)	51.9 (3.08)	64.3 (3.09)	68.4 (3.76)
1901	16.3 (2.58)	12.3 (1.04)	26.5 (5.26)	44.3 (4.58)	54.2 (5.16)	65.0 (2.96)	70.3 (4.13)
1902	15.3 (2.24)	18.1 (2.76)	34.3 (5.36)	43.4 (3.75)	52.1 (4.31)	59.7 (5.00)	65.5 (5.31)
1903	17.4 (3.49)	20.8 (3.91)	37.4 (6.50)	41.8 (1.56)	56.4 (0.41)	59.5 (6.26)	65.8 (4.53)
1904	9.1 (2.85)	11.1 (1.73)	25.7 (2.45)	38.8 (3.97)	58.1 (3.56)	62.7 (2.93)	66.6 (4.06)
1905	12.6 (2.85)	11.4 (1.50)	25.8 (2.62)	40.7 (2.26)	52.8 (2.38)	60.5 (5.18)	68.0 (5.41)
1906	25.2 (2.36)	17.7 (2.75)	22.0 (3.41)	40.1 (1.67)	52.1 (4.71)	62.6 (5.20)	66.7 (3.21)
1907	14.3 (2.17)	9.7 (1.08)	27.8 (2.06)	36.4 (3.59)	47.5 (2.86)	61.3 (4.55)	66.4 (3.95)
1908	16.5 (2.31)	13.4 (3.96)	27.3 (2.53)	38.1 (3.00)	55.7 (5.02)	62.7 (3.17)	69.7 (2.96)
1909	18.6 (4.48)	20.1 (5.13)	25.2 (2.31)	39.4 (3.17)	51.5 (5.14)	62.9 (3.45)	64.8 (2.51)
1910	18.7 (4.21)	15.5 (4.14)	33.7 (1.21)	46.6 (2.65)	52.4 (4.34)	60.6 (3.31)	67.8 (2.35)
1911	18.5 (2.23)	14.3 (2.50)	22.8 (3.73)	37.9 (1.13)	60.4 (1.44)	60.6 (3.30)	69.9 (3.47)
1912	8.3 (2.46)	15.6 (2.12)	21.9 (4.29)	40.4 (2.98)	53.7 (7.31)	58.7 (2.81)	67.0 (2.75)
1913	26.9 (4.16)	15.9 (1.80)	32.1 (7.37)	43.6 (2.08)	50.9 (3.96)	60.6 (1.61)	66.9 (3.36)
1914	14.1 (2.24)	9.7 (1.63)	27.0 (3.87)	37.2 (5.38)	55.0 (1.06)	60.6 (3.28)	65.7 (4.06)
1915	19.6 (3.53)	23.5 (4.64)	24.6 (0.38)	47.2 (1.95)	50.2 (1.94)	61.6 (2.71)	66.8 (7.51)
1916	20.8 (2.15)	15.3 (3.25)	19.5 (2.84)	41.2 (2.82)	52.7 (3.92)	59.3 (5.15)	70.7 (3.23)
1917	16.2 (3.48)	12.7 (1.59)	25.9 (3.15)	38.7 (2.32)	45.4 (2.99)	62.0 (4.92)	69.0 (3.07)
1918	6.0 (2.58)	14.2 (2.68)	26.9 (2.10)	41.8 (2.06)	58.5 (4.23)	58.8 (4.34)	67.8 (2.78)
1919	21.5 (1.92)	20.5 (1.81)	30.2 (5.03)	40.0 (2.40)	53.5 (5.13)	65.7 (3.03)	67.9 (2.40)
1920	6.8 (2.02)	16.2 (3.31)	29.1 (4.12)	39.0 (6.35)	51.3 (1.87)	61.6 (3.77)	64.9 (4.50)
1921	19.4 (1.31)	20.9 (2.26)	35.8 (4.34)	48.0 (3.47)	55.3 (2.06)	62.2 (3.39)	72.8 (3.83)
1922	12.3 (2.29)	18.8 (3.07)	28.8 (5.03)	41.1 (4.20)	54.6 (3.47)	64.0 (8.95)	67.2 (1.98)
1923	11.7 (4.92)	10.3 (1.49)	21.9 (2.81)	39.3 (4.31)	50.3 (2.79)	62.9 (3.16)	63.9 (3.43)
1924	18.7 (4.08)	11.7 (2.25)	28.2 (1.09)	38.2 (4.84)	49.0 (4.25)	60.1 (2.77)	66.8 (3.54)
1925	10.8 (3.20)	24.8 (3.22)	32.2 (4.80)	43.1 (2.47)	48.9 (3.03)	64.3 (5.56)	65.3 (4.89)
1926	17.8 (2.77)	13.9 (3.39)	20.0 (2.78)	34.1 (3.39)	50.2 (1.48)	58.0 (4.71)	65.5 (3.83)
1927	18.4 (2.49)	22.0 (2.64)	31.3 (1.85)	40.3 (1.30)	46.6 (4.05)	57.8 (2.44)	66.9 (5.00)
1928	19.6 (3.53)	18.4 (2.66)	26.2 (3.24)	38.4 (3.75)	51.9 (4.47)	59.8 (4.72)	68.5 (3.65)
1929	15.9 (3.49)	17.7 (3.07)	31.2 (4.42)	41.4 (6.06)	53.3 (4.55)	63.1 (5.39)	66.1 (2.13)
1930	18.9 (3.29)	20.4 (1.24)	27.5 (5.56)	39.3 (1.58)	54.4 (4.93)	66.4 (5.17)	66.1 (3.78)
1931	16.2 (2.11)	18.9 (1.50)	32.0 (2.06)	44.3 (2.61)	55.0 (3.95)	62.6 (3.95)	71.0 (6.91)
1932	28.8 (3.40)	21.1 (1.55)	24.8 (3.05)	39.7 (3.01)	54.4 (2.15)	62.4 (2.55)	65.1 (5.77)
1933	28.3 (1.62)	23.7 (2.33)	26.4 (3.27)	42.6 (5.06)	55.9 (2.36)	65.2 (1.90)	68.3 (2.69)
1934	17.6 (2.91)	5.8 (1.62)	26.6 (2.04)	43.4 (4.16)	56.2 (1.86)	64.9 (4.30)	68.8 (3.31)
1935	12.8 (4.29)	17.7 (1.79)	29.7 (1.31)	41.2 (2.40)	49.6 (2.35)	63.4 (5.59)	70.1 (4.72)
1936	15.1 (4.47)	12.7 (1.76)	36.4 (5.93)	40.2 (4.23)	57.3 (3.28)	63.7 (1.89)	66.2 (3.83)
1937	27.4 (3.24)	24.7 (2.36)	23.7 (2.83)	40.8 (2.45)	56.1 (5.38)	63.5 (4.34)	69.5 (3.78)
1938	16.1 (3.13)	22.3 (2.05)	31.0 (1.92)	45.2 (2.46)	53.0 (2.64)	65.1 (3.21)	68.5 (5.84)
1939	17.3 (2.29)	20.5 (3.22)	24.0 (2.66)	37.7 (4.64)	54.8 (2.41)	63.1 (4.22)	67.7 (2.71)
1940	10.8 (1.07)	17.4 (2.09)	24.6 (4.11)	38.2 (3.37)	55.6 (5.60)	62.3 (3.89)	67.8 (3.73)
1941	14.3 (1.61)	19.8 (1.47)	23.8 (1.97)	47.6 (0.72)	54.8 (2.45)	64.7 (3.14)	68.9 (5.13)
1942	17.6 (1.86)	16.1 (2.15)	34.7 (4.19)	44.9 (2.97)	58.3 (3.20)	63.8 (5.20)	67.3 (2.86)
1943	13.1 (1.63)	19.4 (2.06)	26.5 (2.15)	35.3 (3.42)	52.9 (4.88)	65.5 (5.18)	67.9 (4.24)
1944	19.2 (1.65)	17.0 (2.47)	24.6 (2.48)	37.5 (3.17)	59.5 (2.07)	63.2 (4.64)	68.6 (4.33)
1945	10.0 (3.67)	20.6 (2.18)	38.7 (1.84)	48.2 (4.88)	51.1 (6.02)	62.5 (4.11)	67.8 (5.77)

THE VERMONT WEATHER BOOK

AUG.	SEP.	OCT.	NOV.	DEC.	ANN.	
66.7 (5.54)	60.0 (2.82)	43.0 (1.52)	36.7 (5.51)	27.0 (4.56)	43.9 (39.08)	1895
66.3 (4.17)	56.0 (4.79)	44.0 (2.65)	41.3 (3.92)	21.8 (1.04)	43.1 (39.90)	1896
64.0 (3.93)	56.8 (2.05)	47.7 (1.18)	33.1 (6.96)	23.5 (5.25)	44.4 (57.01)	1897
67.6 (6.81)	60.3 (4.43)	48.5 (4.67)	33.6 (3.88)	22.2 (2.46)	44.4 (48.16)	1898
66.2 (2.33)	56.2 (4.88)	49.3 (1.91)	33.5 (2.56)	25.9 (2.96)	43.2 (38.84)	1899
66.8 (5.44)	60.0 (2.58)	52.0 (2.49)	35.9 (6.15)	21.8 (2.28)	43.3 (45.62)	1900
66.8 (4.94)	58.7 (3.05)	47.1 (1.94)	29.1 (2.40)	22.0 (5.49)	42.7 (43.53)	1901
62.9 (4.85)	58.8 (4.36)	45.2 (3.93)	38.6 (1.42)	17.3 (4.24)	42.6 (47.53)	1902
59.0 (5.20)	58.6 (1.23)	46.5 (2.96)	31.2 (1.75)	16.4 (2.87)	42.6 (40.67)	1903
63.2 (5.29)	54.0 (6.43)	42.9 (2.60)	27.7 (1.20)	12.7 (2.19)	39.4 (39.26)	1904
63.1 (5.20)	56.5 (6.08)	46.8 (1.84)	31.5 (2.55)	23.7 (3.12)	41.1 (40.99)	1905
67.5 (3.78)	58.6 (2.75)	46.6 (2.93)	32.3 (2.53)	16.0 (3.20)	42.3 (38.50)	1906
61.8 (1.96)	57.3 (7.54)	41.0 (4.22)	32.4 (3.95)	24.8 (4.01)	40.1 (41.94)	1907
63.2 (4.86)	60.1 (0.96)	48.7 (1.72)	34.5 (1.39)	19.6 (2.74)	42.5 (34.62)	1908
63.4 (4.13)	56.5 (4.64)	44.2 (1.44)	45.4 (2.12)	18.1 (1.87)	42.5 (40.39)	1909
63.2 (4.84)	56.1 (5.00)	43.0 (1.62)	32.4 (2.89)	13.5 (2.29)	42.0 (38.85)	1910
65.5 (4.92)	55.1 (4.20)	44.8 (3.98)	32.2 (2.02)	28.7 (2.89)	42.6 (35.81)	1911
60.1 (5.23)	56.2 (5.78)	48.5 (3.66)	35.8 (3.72)	26.6 (3.22)	41.1 (46.33)	1912
64.2 (2.03)	51.7 (2.29)	47.2 (4.43)	37.8 (1.69)	23.2 (3.01)	43.4 (37.79)	1913
64.8 (5.84)	55.9 (2.34)	49.4 (1.30)	31.7 (2.62)	18.8 (2.05)	40.8 (35.67)	1914
64.5 (6.15)	61.4 (2.56)	48.6 (2.22)	36.1 (2.51)	22.7 (4.37)	43.9 (40.47)	1915
68.0 (3.27)	57.4 (5.28)	47.1 (1.76)	32.4 (3.82)	21.6 (2.69)	42.2 (40.18)	1916
67.3 (6.10)	55.0 (1.82)	43.7 (5.56)	28.9 (1.03)	9.4 (2.35)	39.5 (38.38)	1917
66.1 (4.34)	53.7 (7.52)	48.1 (4.32)	35.7 (3.19)	24.2 (2.81)	41.8 (42.95)	1918
62.7 (4.68)	57.1 (4.89)	47.4 (4.10)	33.7 (4.04)	16.2 (1.59)	43.0 (41.02)	1919
68.4 (4.15)	59.1 (5.92)	52.6 (2.22)	31.6 (4.86)	24.3 (5.89)	42.1 (48.98)	1920
64.5 (4.08)	60.9 (2.36)	46.9 (2.64)	31.7 (6.24)	17.5 (2.06)	44.7 (38.04)	1921
64.9 (6.13)	59.3 (2.29)	45.5 (1.93)	34.9 (1.87)	20.1 (2.63)	42.6 (43.84)	1922
62.3 (3.04)	58.9 (3.85)	46.7 (3.75)	35.1 (4.24)	28.9 (3.63)	41.0 (41.42)	1923
65.0 (4.73)	55.5 (6.99)	46.5 (0.66)	35.1 (3.38)	17.8 (1.81)	41.0 (40.39)	1924
64.3 (2.73)	57.2 (5.74)	39.2 (3.87)	33.7 (4.29)	21.5 (2.04)	42.1 (45.84)	1925
64.7 (4.25)	56.0 (3.20)	44.3 (4.86)	34.2 (4.37)	15.7 (2.07)	39.5 (41.10)	1926
61.8 (5.54)	57.6 (1.71)	50.0 (5.28)	38.8 (11.85)	24.3 (5.68)	43.0 (49.83)	1927
69.4 (6.02)	54.7 (3.60)	48.2 (2.20)	35.1 (2.30)	27.1 (1.79)	43.1 (42.02)	1928
62.5 (3.44)	59.9 (3.23)	45.4 (2.35)	35.5 (2.72)	20.9 (3.87)	42.7 (44.72)	1929
64.4 (2.66)	60.7 (2.50)	46.5 (1.96)	36.7 (3.13)	23.2 (1.71)	43.7 (37.51)	1930
65.1 (2.48)	60.9 (5.03)	50.0 (2.91)	41.5 (1.78)	25.0 (2.74)	45.2 (38.03)	1931
67.1 (3.69)	59.3 (2.86)	50.0 (4.44)	32.7 (5.14)	27.9 (1.37)	44.4 (38.98)	1932
65.9 (5.03)	59.6 (3.20)	45.8 (3.28)	28.0 (1.70)	16.6 (2.87)	43.9 (35.31)	1933
62.9 (1.83)	62.9 (4.61)	44.3 (1.81)	38.6 (3.16)	18.3 (2.85)	42.5 (34.46)	1934
66.4 (2.79)	55.5 (4.05)	47.9 (1.59)	38.7 (4.34)	17.5 (1.18)	42.5 (36.40)	1935
64.8 (4.23)	58.4 (2.93)	47.0 (5.16)	30.3 (2.64)	25.0 (3.70)	43.1 (44.05)	1936
71.1 (3.57)	58.3 (2.35)	46.6 (4.50)	36.1 (3.37)	21.0 (2.35)	44.9 (40.52)	1937
68.7 (4.50)	55.0 (8.11)	49.7 (1.61)	37.4 (2.48)	24.9 (3.80)	44.7 (41.75)	1938
69.3 (3.22)	58.0 (3.66)	46.8 (4.08)	31.6 (0.82)	22.6 (2.47)	42.8 (36.40)	1939
64.8 (1.86)	57.0 (4.64)	43.6 (1.67)	35.1 (4.64)	23.0 (3.32)	41.7 (39.99)	1940
62.7 (2.54)	59.0 (2.25)	46.4 (3.71)	38.0 (2.55)	25.1 (2.67)	43.8 (30.21)	1941
65.7 (2.35)	59.0 (4.65)	49.1 (3.23)	35.4 (3.51)	18.0 (3.23)	44.2 (39.40)	1942
64.4 (5.49)	56.0 (2.25)	47.3 (4.47)	33.7 (4.95)	16.5 (0.99)	41.5 (41.71)	1943
68.9 (1.86)	59.7 (4.94)	45.6 (3.37)	35.1 (2.62)	18.2 (1.99)	43.1 (35.59)	1944
66.0 (1.89)	60.8 (6.73)	45.4 (5.23)	35.9 (4.35)	17.6 (2.17)	43.7 (48.84)	1945

	JAN.	FEB.	MAR.	APR.	MAY	JUNE	JULY
1946	17.1 (2.22)	15.7 (2.48)	39.5 (1.36)	40.7 (2.30)	52.2 (4.40)	62.4 (2.89)	66.4 (3.26)
1947	20.0 (3.50)	17.3 (2.26)	29.0 (2.88)	39.8 (2.31)	53.2 (5.97)	62.1 (6.07)	70.2 (6.23)
1948	12.0 (2.20)	14.4 (1.93)	28.2 (2.60)	43.5 (3.11)	52.4 (4.37)	61.5 (3.66)	68.2 (3.63)
1949	24.8 (2.94)	24.1 (2.21)	30.6 (2.07)	44.1 (2.98)	55.1 (2.80)	68.4 (2.29)	71.0 (4.31)
1950	25.4 (3.54)	16.3 (2.29)	22.8 (3.01)	39.1 (2.24)	54.7 (1.82)	64.1 (4.05)	67.0 (2.97)
1951	21.3 (2.25)	23.3 (4.26)	31.2 (3.42)	43.6 (4.04)	54.5 (2.44)	62.3 (3.17)	67.7 (5.25)
1952	20.3 (2.62)	22.7 (2.76)	29.6 (2.13)	44.8 (3.20)	50.3 (3.94)	64.0 (6.30)	70.6 (2.57)
1953	24.0 (3.52)	24.7 (1.92)	32.8 (4.41)	43.4 (3.94)	56.6 (4.54)	64.7 (1.88)	68.8 (2.48)
1954	13.8 (2.81)	26.3 (3.58)	29.2 (3.70)	43.3 (4.01)	52.8 (4.85)	63.9 (4.77)	65.2 (3.16)
1955	15.3 (0.97)	20.7 (3.44)	28.2 (3.78)	44.9 (2.60)	58.2 (2.65)	64.0 (3.26)	71.5 (2.20)
1956	20.3 (2.29)	22.1 (2.39)	22.9 (3.87)	38.6 (2.98)	48.9 (3.86)	63.4 (2.68)	64.7 (3.51)
1957	10.0 (1.89)	23.3 (1.19)	30.2 (1.27)	43.7 (2.02)	52.9 (3.33)	66.1 (4.59)	65.7 (4.47)
1958	19.6 (4.69)	12.8 (2.43)	32.3 (1.77)	44.0 (2.92)	50.7 (2.95)	58.1 (2.99)	67.6 (5.21)
1959	16.3 (3.73)	13.1 (2.37)	26.8 (2.28)	43.4 (1.91)	56.9 (1.96)	63.5 (3.59)	70.1 (1.83)
1960	17.8 (2.53)	5.20 (2.93)	23.2 (1.86)	44.0 (4.34)	58.9 (3.34)	63.2 (4.07)	65.2 (4.04)
1961	10.0 (1.21)	20.5 (2.39)	29.2 (1.97)	40.4 (4.16)	51.9 (3.32)	62.9 (4.01)	67.4 (4.05)
1962	16.8 (2.06)	14.7 (2.46)	30.6 (1.62)	42.4 (3.71)	54.2 (2.92)	64.3 (2.48)	63.7 (5.15)
1963	17.1 (2.32)	13.1 (2.00)	29.0 (2.61)	40.8 (2.82)	52.9 (2.89)	64.1 (2.40)	69.0 (2.77)
1964	20.7 (3.30)	17.9 (1.21)	31.2 (3.82)	41.5 (2.71)	57.8 (3.28)	62.1 (1.91)	69.2 (3.24)
1965	14.2 (1.37)	18.0 (2.08)	27.8 (0.87)	38.9 (2.33)	56.4 (0.98)	61.6 (3.02)	63.9 (2.92)
1966	16.8 (2.10)	19.3 (2.32)	30.8 (3.12)	40.1 (1.13)	50.4 (3.11)	63.8 (3.12)	67.3 (3.36)
1967	23.6 (1.54)	12.9 (1.95)	25.0 (1.10)	40.3 (3.80)	46.9 (4.53)	66.4 (3.34)	69.0 (4.03)
1968	10.1 (1.62)	13.8 (1.04)	31.9 (3.82)	47.0 (3.29)	52.1 (3.59)	61.7 (4.33)	68.7 (2.68)
1969	17.8 (2.24)	20.9 (2.59)	27.3 (1.77)	42.6 (3.82)	52.3 (3.96)	63.7 (3.98)	66.8 (4.51)
1970	6.5 (0.76)	18.8 (2.61)	28.0 (2.16)	42.1 (3.60)	55.5 (3.35)	63.4 (3.41)	69.3 (2.10)
1971	10.6 (1.72)	20.3 (3.53)	25.9 (3.42)	37.8 (2.05)	53.5 (3.19)	63.3 (2.20)	67.1 (4.44)
1972	18.9 (1.93)	16.3 (2.65)	26.1 (3.98)	36.2 (2.48)	55.4 (3.86)	62.6 (5.16)	68.1 (5.59)
1973	21.3 (3.02)	16.7 (1.87)	36.8 (2.62)	44.5 (4.23)	52.9 (5.81)	65.4 (7.67)	69.2 (3.96)
1974	19.8 (2.70)	16.9 (2.59)	28.3 (3.74)	43.2 (3.74)	51.1 (5.18)	63.8 (3.39)	67.9 (5.03)
1975	22.1 (2.93)	20.5 (2.55)	26.9 (3.10)	36.5 (2.66)	60.2 (1.49)	63.7 (3.63)	71.1 (3.98)
1976	11.2 (4.19)	23.7 (3.49)	30.7 (3.18)	45.5 (3.07)	52.7 (5.49)	67.3 (4.29)	66.0 (5.37)
1977	11.0 (2.18)	20.8 (2.07)	36.2 (3.85)	43.2 (3.86)	57.3 (1.23)	61.7 (3.77)	66.9 (2.24)
1978	15.4 (5.87)	11.3 (0.71)	25.3 (2.68)	38.4 (2.66)	57.2 (2.17)	62.8 (5.51)	67.2 (3.26)
1979	18.1 (6.41)	9.0 (1.34)	34.9 (2.67)	42.3 (3.43)	56.5 (4.88)	63.0 (1.67)	69.4 (2.90)
1980	20.0 (1.32)	16.1 (0.84)	28.9 (3.54)	43.5 (2.95)	55.2 (1.74)	60.5 (2.81)	68.5 (3.76)
1981	8.0 (.58)	29.4 (6.39)	31.2 (1.13)	45.1 (3.09)	56.0 (4.07)	63.5 (3.61)	68.3 (4.15)
1982	7.7 (2.62)	18.5 (2.71)	28.0 (2.68)	39.7 (3.17)	56.6 (1.58)	60.4 (6.30)	67.5 (2.36)
1983	19.4 (3.64)	21.4 (2.13)	31.5 (4.08)	41.1 (6.10)	51.7 (6.76)	64.0 (2.41)	68.2 (3.04)
1984	14.5 (1.37)	27.8 (3.05)	22.1 (2.99)	43.1 (3.05)	51.4 (8.47)	63.8 (2.79)	67.3 (5.39)
1985	12.2 (1.65)	21.1 (2.11)	30.1 (2.85)	42.0 (2.34)	54.1 (3.39)	58.9 (3.81)	66.6 (3.19)
1986	17.8 (4.42)	16.3 (2.16)	31.3 (3.69)	47.0 (1.63)	56.6 (4.32)	59.9 (4.78)	65.7 (5.26)
1987	15.9 (3.14)	13.9 (0.52)	31.3 (1.87)	46.3 (2.85)	53.6 (2.03)	63.2 (5.56)	68.4 (3.85)
1988	16.4 (1.52)	19.1 (2.61)	27.9 (1.97)	42.2 (3.59)	55.7 (2.44)	60.5 (2.22)	69.8 (4.11)
1989	20.9 (1.30)	17.5 (1.69)	26.4 (2.89)	38.1 (2.64)	56.0 (5.48)	63.4 (4.86)	67.7 (3.47)
1990	26.8 (3.58)	21.2 (3.10)	30.8 (3.00)	42.9 (3.78)	50.3 (5.26)	62.8 (4.31)	67.1 (3.99)
1991	16.0 (2.46)	23.7 (1.27)	31.5 (3.37)	45.5 (3.24)	56.5 (3.62)	62.9 (2.66)	66.4 (2.73)
1992	16.8 (2.50)	18.9 (2.22)	24.1 (3.10)	38.9 (3.03)	53.1 (2.35)	60.6 (3.02)	62.9 (4.23)
1993	20.1 (3.12)	9.5 (2.72)	25.5 (3.32)	42.2 (4.05)	53.4 (2.22)	62.7 (4.30)	69.3 (3.63)
1994	5.6 (3.76)	12.7 (1.49)	27.3 (4.06)	41.2 (4.18)	51.0 (4.16)	64.8 (4.04)	69.6 (4.66)
1995	24.4 (2.99)	16.0 (2.24)	32.4 (2.33)	36.9 (2.11)	51.8 (2.55)	64.5 (1.32)	69.5 (4.85)

THE VERMONT WEATHER BOOK

AUG.	SEP.	OCT.	NOV.	DEC.	ANN.	
62.6 (4.86)	60.3 (4.52)	50.4 (3.60)	38.4 (2.82)	24.2 (3.56)	44.2 (38.27)	1946
70.0 (1.97)	59.5 (2.34)	53.6 (1.02)	32.5 (4.21)	17.4 (1.78)	43.7 (40.54)	1947
67.2 (3.66)	59.0 (1.26)	45.6 (2.46)	42.5 (5.56)	26.6 (3.93)	43.4 (38.37)	1948
67.6 (4.08)	56.5 (3.47)	51.6 (2.36)	32.6 (2.79)	26.2 (2.33)	46.0 (34.63)	1949
64.5 (4.92)	54.3 (2.51)	49.4 (2.18)	39.8 (4.90)	24.6 (3.20)	43.5 (37.63)	1950
64.4 (3.11)	58.2 (3.42)	48.9 (2.84)	31.1 (4.42)	23.0 (3.63)	44.1 (42.25)	1951
66.5 (2.79)	59.5 (3.31)	44.3 (2.63)	37.3 (1.54)	25.8 (4.15)	44.6 (37.94)	1952
65.0 (3.69)	58.6 (2.88)	48.6 (2.82)	40.7 (1.90)	30.2 (3.37)	46.5 (37.35)	1953
63.0 (3.92)	56.3 (5.31)	51.1 (2.81)	36.9 (4.32)	23.1 (3.43)	43.7 (46.67)	1954
69.7 (7.59)	56.8 (2.21)	48.7 (4.68)	34.0 (2.82)	15.3 (1.04)	43.9 (37.24)	1955
64.5 (2.49)	54.5 (4.46)	47.9 (1.60)	37.0 (2.71)	26.2 (3.00)	42.6 (35.84)	1956
61.9 (1.28)	59.0 (3.38)	47.0 (1.93)	37.8 (3.81)	28.2 (4.67)	43.8 (33.83)	1957
65.6 (2.80)	58.1 (3.70)	46.1 (3.93)	36.9 (2.68)	12.2 (1.22)	42.0 (37.29)	1958
69.2 (4.74)	61.5 (2.12)	47.8 (6.92)	35.3 (5.80)	25.6 (3.28)	44.1 (41.03)	1959
65.0 (1.98)	58.9 (5.06)	45.5 (3.73)	39.0 (2.24)	18.6 (1.27)	43.7 (37.39)	1960
66.3 (3.66)	65.7 (1.94)	50.4 (1.97)	36.9 (3.07)	24.2 (2.47)	43.8 (34.22)	1961
65.3 (3.51)	56.0 (3.35)	46.6 (5.12)	32.5 (2.68)	19.6 (2.36)	42.2 (37.42)	1962
62.3 (4.67)	54.0 (1.88)	52.1 (0.52)	40.8 (5.06)	13.7 (1.73)	42.4 (31.67)	1963
61.1 (4.54)	55.8 (1.05)	45.4 (1.86)	35.7 (2.58)	23.4 (2.92)	43.5 (32.42)	1964
65.5 (4.67)	58.3 (4.70)	46.2 (3.52)	33.2 (4.04)	26.2 (1.91)	42.5 (32.41)	1965
65.8 (5.10)	55.3 (3.73)	46.1 (2.32)	39.9 (2.33)	23.6 (2.96)	43.3 (34.70)	1966
66.2 (3.61)	57.4 (2.49)	48.7 (3.85)	31.6 (2.51)	24.7 (3.82)	42.7 (36.57)	1967
63.7 (2.40)	61.3 (2.87)	50.1 (2.76)	33.2 (5.17)	19.1 (4.61)	42.7 (38.18)	1968
67.9 (4.18)	59.2 (1.98)	46.6 (2.18)	37.2 (5.65)	20.5 (5.50)	43.6 (42.36)	1969
67.2 (3.85)	59.2 (4.00)	50.5 (2.59)	38.5 (2.40)	17.0 (3.84)	43.0 (34.67)	1970
65.6 (6.38)	61.7 (2.10)	53.1 (2.27)	32.6 (3.65)	23.6 (3.05)	42.9 (38.00)	1971
64.2 (2.87)	58.0 (1.63)	42.5 (3.26)	32.5 (5.25)	23.0 (5.45)	42.0 (44.11)	1972
70.3 (3.32)	57.3 (3.99)	48.5 (2.47)	36.2 (3.12)	26.4 (7.51)	45.5 (49.59)	1973
66.4 (3.69)	57.0 (4.86)	41.7 (1.19)	35.6 (4.08)	27.2 (2.66)	43.2 (42.85)	1974
66.6 (4.90)	56.1 (5.99)	49.0 (4.88)	40.6 (4.30)	20.1 (3.24)	44.4 (43.65)	1975
64.9 (5.59)	56.0 (3.95)	43.4 (5.33)	32.1 (1.94)	15.9 (2.25)	42.4 (48.14)	1976
65.5 (5.10)	57.4 (5.96)	45.6 (5.80)	38.6 (4.18)	21.1 (4.30)	43.8 (44.54)	1977
66.0 (3.19)	53.7 (2.11)	44.7 (3.90)	34.2 (1.62)	22.8 (3.14)	41.6 (36.82)	1978
64.7 (4.49)	56.6 (4.00)	46.6 (3.19)	40.6 (3.34)	26.7 (1.94)	44.0 (40.26)	1979
67.5 (3.56)	57.3 (4.12)	44.4 (3.06)	31.4 (3.44)	14.6 (1.82)	42.3 (32.96)	1980
64.4 (4.25)	57.0 (5.86)	42.8 (5.42)	35.6 (2.30)	24.1 (3.29)	43.8 (44.14)	1981
62.6 (3.34)	58.8 (2.74)	47.0 (1.89)	39.2 (3.63)	28.9 (1.89)	42.9 (34.91)	1982
66.4 (5.93)	60.2 (2.31)	46.4 (3.38)	35.8 (7.02)	19.6 (6.27)	43.8 (53.07)	1983
68.4 (3.16)	55.1 (2.36)	48.7 (2.56)	35.3 (3.72)	28.2 (3.86)	43.8 (42.77)	1984
63.7 (4.02)	57.9 (4.31)	46.7 (3.52)	35.1 (4.40)	18.7 (2.23)	42.3 (37.82)	1985
63.7 (5.69)	55.2 (3.67)	45.4 (2.51)	32.2 (3.87)	25.2 (2.95)	43.0 (44.95)	1986
62.9 (2.91)	56.6 (5.43)	43.6 (5.36)	34.1 (3.03)	26.1 (2.05)	43.0 (38.60)	1987
67.4 (5.59)	54.8 (2.16)	41.6 (2.37)	35.9 (5.61)	19.8 (1.63)	42.6 (35.82)	1988
64.2 (6.21)	57.7 (5.15)	47.0 (4.52)	33.3 (3.82)	5.00 (1.36)	41.4 (43.39)	1989
66.5 (7.05)	56.1 (2.81)	48.1 (6.74)	36.1 (3.79)	27.1 (5.41)	44.7 (52.82)	1990
66.7 (6.22)	54.4 (5.18)	47.8 (4.87)	35.4 (2.53)	22.8 (2.87)	44.1 (41.02)	1991
63.2 (3.81)	56.6 (4.11)	41.5 (2.96)	33.0 (4.32)	24.4 (2.05)	41.2 (37.70)	1992
68.1 (3.35)	57.0 (4.56)	43.5 (3.36)	34.6 (2.64)	23.6 (2.09)	42.5 (39.36)	1993
63.3 (4.95)	55.6 (3.80)	46.6 (.97)	39.9 (2.19)	27.8 (2.71)	42.1 (40.97)	1994
65.8 (4.93)	53.5 (3.57)	51.0 (8.00)	32.2 (4.08)	19.8 (2.85)	43.2 (41.82)	1995

* The 1895–1982 data are from "Statewide Average Climate History, Vermont 1895–1982," Historical Climatology Series 6-1 NCDC Sept. 1983. The 1983–1995 data are from the National Climatic Data Center, Asheville, North Carolina, updated for changes in observational times at cooperative stations.

287

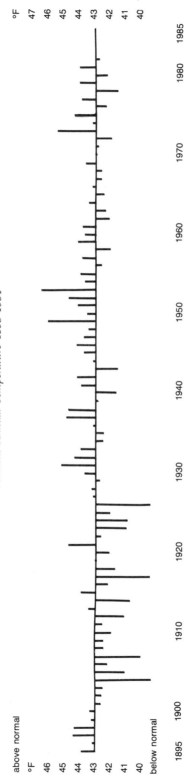

Vermont Annual Temperature 1895–1984

THE VERMONT WEATHER BOOK

Monthly Temperature Extremes

Month	Maximum	Day/Year	Minimum	Day/Year
January	70° Dorset	4 1950	−44° Enosburg	19 1904
			−44° Bloomfield	14 1914
			−44° Bloomfield	31 1934
February	65° Rutland	26 1957	−46° East Barnet	16 1943
March	84° Bennington	29 1945	−36° Bloomfield	8 1913
	84° Burlington	29 1946		
April	97° Vernon	19 1976	−12° Bloomfield	1 1923
May	95° Cornwall	30 1929	14° Bloomfield	3 1946
	95° Bloomfield	30 1929		
	95° Bellows Falls	20 1962		
June	101° St. Johnsbury	4 1919	22° Somerset	22 1919
July	105° Vernon	4 1911	30° Somerset	12 1926
August	102° Bellows Falls	6 1955	25° South Londonderry	31 1965
September	100° Bellows Falls	3 1953	15° Dorset	27 1947
	100° Vernon	3 1953		
October	92° St. Johnsbury*	10 1909	5° Morrisville	19 & 21 1972
	91° Vernon	8 1963		
November	81° Bellows Falls	3 1950	−19° White River Junction	26 1938
December	72° Enosburg Falls	5 1941	−50° Bloomfield	30 1933

* A questionable figure since no other stations had a maximum within five degrees.

TEMPERATURE AND PRECIPITATION RECORDS

Coldest Months in Vermont Since 1895 (average monthly temperatures)

1.	December 1989	5.0°F	5. January 1970	6.5°F
2.	February 1934	5.8°F	6. January 1920	6.8°F
3.	January 1994	5.9°F	7. January 1982	7.3°F
4.	January 1918	6.0°F	8. January 1981	7.7°F

9.	January 1912	8.3°F
10.	February 1979	9.0°F
11.	January 1904	9.1°F
12.	December 1917	9.4°F

Source: Dr. Keith L. Eggleston, Northeast Regional Climate Center, Cornell University.

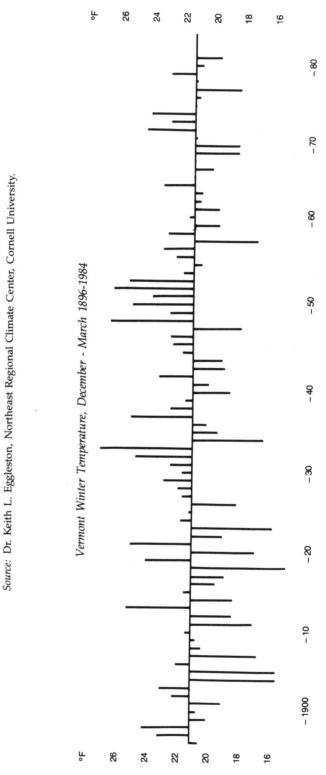

Vermont Winter Temperature, December – March 1896-1984

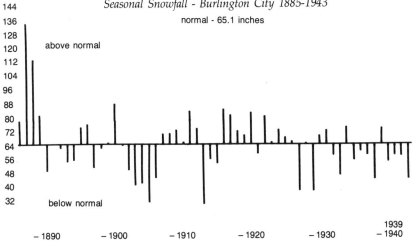

Seasonal Snowfall - Burlington City 1885-1943
normal - 65.1 inches

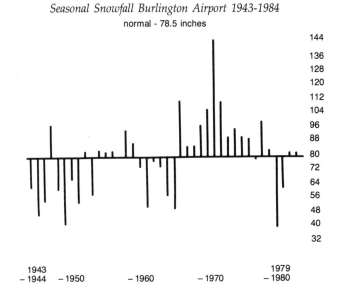

Seasonal Snowfall Burlington Airport 1943-1984
normal - 78.5 inches

291

TEMPERATURE AND PRECIPITATION RECORDS

Normals of Temperature and Precipitation, 1961–1990, in Degrees Fahrenheit (Inches)*

Station & County	JAN.	FEB.	MAR.	APR.	MAY	JUN.	JUL.	AUG.	SEP.	OCT.	NOV.	DEC.	ANN.	ELEVATION (FT.)
Bellows Falls (Windham)	18.7 (2.89)	21.3 (2.82)	32.2 (3.13)	44.0 (3.27)	55.8 (3.54)	64.9 (3.14)	70.1 (3.37)	67.9 (3.96)	59.7 (3.28)	48.5 (3.26)	38.0 (3.59)	24.6 (3.51)	45.5 (39.76)	270
Burlington (Chittenden)	16.3 (1.82)	18.2 (1.63)	30.7 (2.23)	43.9 (2.76)	56.3 (3.12)	65.2 (3.47)	70.5 (3.65)	67.9 (4.06)	58.9 (3.30)	47.8 (2.88)	36.8 (3.13)	23.0 (2.42)	44.6 (34.47)	332
Cavendish (Windham)	16.5 (2.99)	18.8 (2.89)	30.0 (3.39)	41.9 (3.79)	54.4 (4.09)	63.0 (3.86)	67.8 (3.62)	65.5 (4.00)	57.5 (3.49)	46.4 (3.56)	35.6 (4.08)	22.1 (3.77)	43.3 (43.53)	800
Chelsea (Orange)	13.0 (2.41)	15.2 (2.11)	26.9 (2.49)	39.7 (2.87)	51.8 (3.49)	60.8 (3.15)	65.7 (3.48)	63.5 (3.77)	55.4 (3.20)	44.6 (3.21)	33.8 (3.47)	19.5 (3.00)	40.8 (36.65)	800
Cornwall (Addison)	18.6 (2.01)	20.9 (1.81)	32.0 (2.30)	44.7 (2.57)	56.8 (3.40)	65.4 (3.09)	70.1 (3.41)	67.8 (4.35)	59.8 (3.22)	48.9 (2.93)	37.6 (3.19)	24.1 (2.72)	45.6 (35.00)	490
Enosburg Falls (Franklin)	15.9 (2.46)	18.0 (2.06)	30.2 (2.73)	42.9 (3.40)	55.5 (3.94)	63.9 (4.13)	68.4 (4.32)	66.0 (5.23)	58.2 (3.96)	47.9 (3.90)	36.2 (4.37)	21.9 (3.26)	43.8 (43.76)	422
Montpelier Airport (Washington)	15.3 (2.17)	17.5 (2.06)	28.3 (2.32)	41.0 (2.42)	53.4 (3.24)	62.0 (3.48)	66.8 (3.13)	64.4 (3.83)	56.4 (2.88)	45.9 (2.84)	34.5 (3.20)	20.8 (2.81)	42.2 (34.38)	1,126
Newport (Orleans)	14.9 (2.46)	17.9 (2.31)	28.9 (2.76)	41.8 (2.94)	54.8 (3.51)	63.6 (4.05)	68.1 (4.01)	65.9 (4.49)	57.7 (3.54)	47.2 (3.32)	35.0 (3.66)	20.7 (3.33)	43.0 (40.38)	766
Readsboro (Bennington)	18.4 (3.49)	20.1 (3.43)	30.2 (3.86)	42.0 (4.32)	53.7 (4.59)	62.2 (4.54)	67.2 (4.08)	65.2 (4.29)	57.5 (3.79)	47.0 (3.80)	36.9 (4.61)	24.1 (4.28)	43.7 (49.08)	1,120
Rutland (Rutland)	20.7 (2.17)	22.9 (1.89)	33.6 (2.29)	45.4 (2.66)	57.4 (3.48)	65.7 (3.57)	70.3 (3.87)	68.3 (4.19)	60.2 (3.56)	49.7 (2.91)	38.9 (3.13)	25.9 (2.71)	46.6 (36.43)	620
St. Johnsbury (Caledonia)	16.4 (2.30)	19.2 (2.12)	30.6 (2.46)	43.1 (2.73)	56.0 (3.26)	64.9 (3.84)	69.4 (3.67)	67.3 (3.98)	59.1 (3.10)	47.9 (3.09)	35.9 (3.45)	21.9 (3.21)	44.3 (37.21)	699
Vernon (Windham)	19.4 (3.27)	22.1 (3.24)	33.1 (3.60)	44.6 (3.83)	56.4 (4.19)	65.3 (3.86)	70.7 (3.70)	68.5 (3.89)	60.6 (3.41)	48.9 (3.50)	38.6 (4.10)	25.4 (3.84)	46.1 (44.43)	226

West Burke (Caledonia)	10.2 (2.59)	12.1 (2.30)	24.2 (2.74)	37.9 (2.89)	50.6 (3.62)	59.8 (4.25)	64.6 (4.28)	62.2 (4.72)	53.9 (3.43)	42.7 (3.39)	31.4 (3.76)	16.8 (3.50)	38.9 (41.47)	900
Woodstock (Windsor)	14.5 (2.91)	17.1 (2.79)	29.1 (2.97)	41.5 (3.26)	53.8 (3.82)	62.7 (3.18)	67.6 (3.29)	65.1 (3.71)	56.9 (3.40)	45.2 (3.49)	35.0 (3.73)	20.8 (3.55)	42.4 (40.10)	750
Statewide	15.8 (2.49)	18.0 (2.29)	29.1 (2.76)	41.5 (3.16)	53.8 (3.69)	62.7 (3.76)	67.5 (3.81)	65.1 (4.38)	57.0 (3.45)	46.2 (3.34)	35.2 (3.72)	21.7 (3.28)	42.8 (40.13)	

* According to the standards of the World Meteorological Society, normals for temperature and precipitation are for thirty-year periods. The normals for 1961–1990 will be employed through 2000, when new data will be calculated and employed for the next decade.

Source: Monthly Normals of Temperature, Precipitation, and Heating and Cooling Degree Days, 1961–90, Vermont. Climatography of the United States, No. 81. NOAA, National Environmental Satellite, Data, and Information Service, National Climatic Center, Asheville, N.C. 28801-2696.

Vermont Annual Precipitation 1895-1984

TEMPERATURE AND PRECIPITATION RECORDS

GLOSSARY

advection – horizontal movement of any atmospheric element; of air, moisture, or heat, for example.

air mass – an extensive body of air whose horizontal distribution of temperature and moisture is nearly uniform.

airstream – a substantial body of air with the same characteristics flowing with the general circulation.

Alberta-type storm – a cyclonic disturbance originating in the lee of the Rocky Mountains of Alberta; frequent in all seasons; in winter, a fast-mover that brings quick changes in local weather.

anticyclone – an atmospheric pressure system characterized by relatively high pressure at its center and winds blowing outward in clockwise fashion; *also called* a high-pressure area or, simply, a high.

Arctic air – an air mass that has been conditioned over very high latitudes; colder and drier than polar air.

atmospheric pressure – the weight per unit area of the total mass of air above a given point; *also called* barometric pressure.

backing wind – a wind that shifts counterclockwise, that is, from northeast to north to northwest.

barometric pressure – **See atmospheric pressure.**

Bermuda High – a semipermanent anticyclonic area overlying the central North Atlantic Ocean. When it extends westward to the Carolinas, it causes the northward movement of warm, humid airstreams from the Gulf of Mexico that lead to heat waves.

blizzard – a severe winter storm accompanied by bitter cold, strong winds, and snow that may be blown along the surface in blinding sheets. In the East, any heavy snowstorm is often called a blizzard.

circulation – the flow pattern of moving air. The *general circulation* is the large-scale flow characteristic of the semipermanent pressure systems, while the *secondary circulation* occurs in the more temporary, migratory high- and low-pressure systems.

coastal storm – a cyclonic, low-pressure system moving along a coastal plain or just offshore. It causes north to northeast winds over the land; along the Atlantic seaboard it is called a northeaster.

cold front – the interface or transition zone between advancing cold air and retreating warm air.

cold outbreak – the descent of very cold air from higher latitudes into the United States, usually following a sharp cold front.

condensation – the process whereby a substance changes from the vapor phase to the liquid or solid phase; the opposite of evaporation.

conduction – transmission of heat by direct contact through a material substance, as distinguished from convection, advection, and radiation.

convection – transfer of heat by movement of material bodies. In meteorology, convection refers particularly to the thermally induced, vertical motion of air.

convergence – a distribution of wind movement that results in an inflow of air into an area such as a low-pressure area.

cooling degree days – The number of degrees by which the mean temperature for the day exceeds 65° for that day: 90° mean = 25 cooling degree days. This number provides a means of estimating relative energy requirements for air conditioning and is arrived at by a method similar to determining degree days.

cyclogenesis – the process leading to the development of a new low-pressure system, or the intensification of a pre-existing one.

cyclone – an atmospheric pressure system characterized by relatively low pressure at its center and winds blowing inward in a counterclockwise fashion (in the Northern Hemisphere); *also known* as a low-pressure system, or, simply, a low.

deepening – the decrease of pressure at the center of a storm system.

degree days – "heating degrees for the day." Degree days are based on the premise that artificial heating is not needed at mean temperatures above 65°F (18°C), and are calculated by taking the mean temperature, that is, half the sum of the day's maximum and minimum temperatures, and subtracting the mean from the base of 65: a 30° mean equals 35 degree days, for example. Degree days are totaled for the month and season to determine a relative value for heat consumption and costs.

depression – an area of low pressure.

dew – liquid water droplets on other objects caused by condensation of water vapor from the air as a result of radiation cooling.

dew point – the temperature at which a parcel of air reaches saturation as it is cooled at constant pressure.

discontinuity – in meteorology, the rapid variation of the gradient of an element, such as the rate of pressure or temperature change at a front.

disturbance – an area of low pressure attended by storm conditions.

divergence – a distribution of wind movement that results in a net outflow of air from an area such as a high-pressure system. *Cf.* **convergence.**

drought – a condition with subnormal precipitation over a substantial period; its severity depends on previous soil moisture conditions and the duration of dry period.

evaporation—the change of a substance from the liquid to the vapor or gaseous stage; the opposite of condensation.

extra-tropical cyclone—an atmospheric disturbance that either originates outside the tropics, or, having left the tropics, has lost the characteristics of a tropical storm.

eye of the storm—a roughly circular area of comparatively light winds and broken clouds found in the center of many tropical storms.

filling—the increase of pressure at the center of a storm system.

fog—a visible aggregate of minute water droplets suspended in the atmosphere near the earth's surface. Can reduce horizontal visibility to near zero; frequent over lakes and in river valleys in early morning.

front—the interface or transition zone between air masses of different densities and characteristics.

frost—ice crystals formed on grass or other objects by the sublimation of water vapor from the air at below-freezing temperatures.

glaze—a sheath of transparent ice resulting from an ice storm.

hail—precipitation in the form of balls or lumps of ice. Usually accompanies a thunderstorm and can cause damage to crops and buildings.

haze—fine dust, salt, pollen, or other particles dispersed through the atmosphere; they reduce visibility and give the atmosphere a whitish appearance; most frequent during autumn and during Indian summer.

high-pressure system—a system characterized by relatively high pressure at the center; usually attended by fair weather. *Also called* a high.

Hudson Bay High—an anticyclone originating over central Canada and possessing cold, dry air that can cause a cold wave; most frequent in late winter and spring.

hurricane—in the Western Hemisphere, a severe storm of tropical origin and having a cyclonic structure with an eye or calm area at its center. Such a storm must attain a wind speed of 73 miles per hour, otherwise it is classified as a tropical storm.

icestorm—precipitation from clouds whose temperatures are above freezing, that freezes on impact with objects whose temperature is below freezing. An icestorm may occur on a mountainside in a transition zone between rain and snow.

instability—a condition of the atmosphere whereby a parcel of air given an initial vertical impulse will tend to continue to move upward. Often forms cumulus clouds and occasionally thunderstorms.

inversion—a reversal of the normal decrease of temperature with increasing altitude; above the inversion line the temperature increases, or decreases less rapidly than before. Causes air pollution in lower layer of atmosphere.

jet stream—a zone of relatively strong winds concentrated within a narrow zone in the upper atmosphere; usually the location of the maximum winds imbedded in the westerlies.

lake-effect mechanism—a turbulent action of the atmosphere over a warm body of water when a cool airstream passes over; may result in rain showers or snow showers, sometimes heavy, on the lee shore of the lake.

low-pressure system. See **cyclone.**

297

mean temperature – the average of any series of temperatures observed over a period of time. The mean daily temperature is the average of twenty-four hourly temperatures. For convenience, it is usually given as the average of the maximum and minimum temperatures for a twenty-four-hour period.

normal – the average value of a meteorological element over a fixed period of years. Current normals are for the thirty-year period, 1951-1980; they are updated every ten years.

occluded front – a composition of two fronts produced when a cold front overtakes a warm front, forcing the warm air aloft.

overrunning – the ascent of warm air over relatively cool air; usually occurs in advance of a warm front.

polar air – an air mass conditioned over the tundra or snow-covered terrain of high latitudes.

polar front – a semipermanent discontinuity separating cold polar easterly winds and relatively warm westerly winds of the middle latitudes.

precipitation – products of condensation that fall as rain, snow, hail, or drizzle.

prefrontal squall line – an unstable line of turbulence preceding a cold front at some distance, often accompanied by showers or thunderstorms.

radiation – the transfer of energy through space without the agency of intervening matter.

radiational cooling – the cooling of the earth's surface and adjacent air (mainly at night) whenever the earth's surface suffers a net loss of heat. The heat radiates to outer space when the sky is clear of clouds.

relative humidity – the ratio of the actual amount of water vapor in a given volume of air to the amount that could be present if the air were saturated, temperature remaining constant. Commonly expressed as a percentage.

ridge – an elongated area of high barometric pressure.

saturation – condition of a parcel of air holding a maximum of water vapor; a 100-percent relative-humidity condition exists.

secondary cold front – a cold front that may form behind the primary cold front and carry an even colder flow of air.

secondary depression – an area of low pressure that forms in a trough to the south or east of the primary storm center.

semipermanent high or low – one of the relatively stationary and stable pressure and wind systems; for example, the Icelandic Low, the Bermuda High.

sleet – small pellets of ice in the form of balls or lumps of ice falling from clouds with below-freezing temperatures; distinguished from glaze, which is a covering of ice on exposed objects; often falls in a transition zone between rain and snow.

smog – a natural fog contaminated by industrial pollutants; often a mixture of smoke and fog.

source region – an area of nearly uniform surface characteristics over which large bodies of air stagnate and acquire a more or less equal horizontal distribution of temperature and moisture.

squall line – a well-marked line of instability ahead of a cold front accompanied by strong gusty winds, turbulence, and often heavy showers.

stationary front – an interface zone between cold and warm air that exhibits little or no movement.

steering—the process whereby the direction of movement of surface pressure systems is influenced by the circulation aloft.

subsidence—the descending motion aloft of a body of air, usually within an anticyclone; causes a spreading-out and warming of the lower layers of the atmosphere.

tendency—the local rate of change of a meteorological element; usually refers to a barometric rise or fall.

tornado—a violently rotating column of air extending downward from a cumulonimbus, or thunderhead, cloud, usually having the appearance of a funnel and often filled with debris.

tropical air—an air mass conditioned over the warm surfaces of tropical seas or land.

tropical storm—a cyclonic storm with a circulatory wind system that originates over tropical oceans. It usually has a warm core and distinctive central eye. According to the National Weather Service classification of a tropical storm, has winds from 39 to 73 mi/h; if higher, it is designated a hurricane.

trough—an elongated area of low pressure.

typhoon—the name applied in the western Pacific Ocean to severe tropical storms; their structures are similar to hurricanes.

veering wind—occurs when winds shift clockwise, that is, from southwest to west to northwest.

warm front—the line of advancing warm air that is displacing cooler air at the surface.

warm sector—the portion of a cyclone containing warm air, usually located in the southeastern sector of the storm system.

waterspout—a funnel-shaped, tornadolike cloud that originates over a body of water.

water vapor—atmospheric moisture in the invisible gaseous phase.

whirlwind—a rapidly whirling, small-scale vortex of air, often seen on hot still days.

wind—air in motion, occurring naturally in the atmosphere and caused by a difference in densities of nearby air parcels; refers to air moving parallel with the surface of the ground.

wind chill factor—a figure derived from the attempted calculation of heat loss from exposed human skin through the combination of particular temperatures and air speeds and involves heat loss from four factors: conduction, convection, radiation, and evaporation. It is based on a formula calculated from Dr. Paul Siple's experimental observations at the Antarctica base Little America of the effect of wind and cold on exposed human flesh. A chart has been constructed for quick determination of the factor if one knows the wind speed and the current temperature.

SOURCES OF CLIMATOLOGICAL DATA

Summaries

Statewide Average Climatic History, Vermont 1895-1982. Historical Climatology Series 6-1. National Climatic Data Center, Asheville, North Carolina. September 1983. 29 pages. Contains monthly and annual temperature and precipitation data for 1895 to 1982.

Climatic Summary of the United States, Section 84, New Hampshire and Vermont. Contains monthly and annual temperature and precipitation for Burlington, 1884-1930. United States Weather Bureau, Department of Agriculture, Washington, D.C. 1933. 28 pages. Precipitation data, monthly and annual, for twenty-seven Vermont stations from the beginning of records to 1930. Also frost data for twelve stations.

Ingram, Robert S., and Samuel C. Wiggans. *Climate of Burlington, Vermont.* Vermont Agricultural Experiment Station, University of Vermont, Burlington, Vermont. October 1968. 50 pages. Contains history of weather observations at Burlington. Also numerous tables and charts summarizing the city's weather elements.

Monthly Normals of Temperature, Precipitation, and Heating and Cooling Degree Days, 1951-1980, Vermont. Climatography of the United States, No. 81. National Climatic Data Center, Asheville, North Carolina. September 1982, 12 pages.

Monthly Averages of Temperature and Precipitation for State Climatic Divisions, 1941-70, New England. Climatography of the United States, No. 85. National Climatic Data Center, Asheville, North Carolina. July 1973. 16 pages. Contains monthly and annual temperature and precipitation averages for Vermont.

Periodicals

Local Climatological Data. Monthly Summary. International Airport, Burlington, Vermont. National Climatic Data Center, Asheville, North Carolina. 2 pages. Daily meteorological data, issued monthly.

Climatological Data. New England Section. National Climatic Data Center, Asheville, North Carolina. Twelve monthly issues and annual summary. Contains temperature and precipitation data for about forty-four Vermont stations. Publication started in 1896, now in volume 97.

Vermont Climate. Monthly. Vermont Climate Office and Northeast Regional Climate Center, 1112 Bradfield Hall, Cornell University, Ithaca, New York 14853. Contains discussion of month's weather, a climatological preview of coming month, and data summary for current month.

Storm Data. Monthly. National Climatic Data Center, Asheville, North Carolina. Contains brief reports on severe storms for all states.